HEALTHY BRAIN, HAPPY LIFE

HEALTHY BRAIN, HAPPY LIFE

A PERSONAL PROGRAM TO
ACTIVATE YOUR BRAIN AND DO EVERYTHING BETTER

WENDY SUZUKI, PhD
WITH BILLIE FITZPATRICK

DEY ST.
AN IMPRINT OF
WILLIAM MORROW *PUBLISHERS*

Illustrations on pages 14, 16, and 192 courtesy of Ashley Halsey.

The figure on page 125 was reproduced with permission from Brickman, A. M., Khan, U. A., Provenzano, F. A., Yeung, L. K., Suzuki, W., Schroeter, H., Wall, M., Sloan, R. P., and Small, S. A. "Enhancing Dentate Gyrus Function with Dietary Flavanols Improves Cognition in Older Adults." *Nature Neuroscience* 17 (2014): 1798–803.

All photographs are courtesy of the author.

Consult your medical professional before beginning any exercise program. The author and publisher are not responsible for any losses or damages that result from the application of the material in this book.

HarperCollins books may be purchased for educational, business, or sales promotional use. For information please e-mail the Special Markets Department at SPsales@harpercollins.com.

FIRST EDITION

Designed by Renato Stanisic

Library of Congress Cataloging-in-Publication Data has been applied for.

ISBN 978-0-06-236678-8

15 16 17 18 19   OV/RRD   10 9 8 7 6 5 4 3 2 1

For Mom and Dad
I love you both

CONTENTS

HEALTHY BRAIN, HAPPY LIFE

INTRODUCTION

One day I woke up and realized I didn't have a life. As an almost-forty-year-old award-winning, world-renowned neuroscientist, I had what many considered *everything*. I had achieved my life-long dream of running my own successful and highly respected neuroscience research lab at New York University and had earned tenure as a professor. These are both feats that are extremely hard to achieve for many, many reasons. Too many of my female friends from graduate school, where the ratio was fifty–fifty male to female, had drifted away from science. The reasons were in some ways common for women in any profession. Their husbands got a job in a place where there wasn't a job for them in science, or they took time off to have kids and found it hard or impossible to come back. They might have become discouraged by the ultra-competitive grant-writing process, or they just got tired of the long hours and low pay and found other outlets for their talent and creativity. The women, like me, who soldiered on in science were few and far between. Specifically, women now make up on average 28 percent of the science faculty at most major U.S. research institutions. The roller coaster–like drop from 50 percent women in graduate school to an average of 28 percent women at the faculty level works like a big flashing neon warning sign to women, saying, "Beware: Life is mighty challenging in this neck of the woods!"

Despite the depressing statistics, I kept moving forward. I published many articles in prestigious scientific journals and won many prizes for my work on the anatomy and physiology underlying memory function in the brain. I was a role model for women scientists and highly respected by my peers. On paper, I had a stellar career and an impeccable track record. And I loved doing science—I really did.

What could possibly be wrong? Well . . . *everything else.*

To be honest, my life was pretty depressing. While I had created a dream career for myself, I had no social life and no boyfriend in sight. I had strained relationships with members of my department and my own lab. When a senior faculty colleague I was teaching with decided he would write exams, grade exams, and set up laboratory exercises at the very last minute, I felt completely powerless to push back. When a student decided (without telling me) to take a chunk of time off from her research with me to take a teaching job, I was incensed. The only way I knew how to relate with the other scientists in my lab was through work—or more accurately, working really hard. I couldn't talk to them about anything else in life because in my mind there *was* nothing else in life. Oh, and did I mention that I was fat too? Twenty pounds overweight to be exact. I felt miserable, and for the first time in my life, completely without direction. I was really good at engaging with science and advancing my career, but it seemed that I was really bad at living. Now don't get me wrong, I loved what I was doing. I was and always have been passionate about science. But could I get by on work alone?

Then I came to a startling realization: I was truly clueless at something, something really important.

What does a woman of science do when she realizes that she is missing out on everything but science?

In my case, I decided to conduct an experiment on myself, which changed the course of my life.

Over these past few years, I have leveraged my twenty years of neuroscience research and taken a wild leap of faith. I ventured beyond the world of science and discovered a whole new universe of health and happiness that ironically led me right back to where I started. And an enormous, almost total transformation happened inside of *me*.

Determined to change my fate, I went from living as a virtual lab rat—an overweight middle-aged woman who had achieved many things in science, but who could not seem to figure out how to also be a healthy, happy woman with both a stimulating career and meaningful relationships. I was at a profound low point, and I knew that the only person who could lift me up was me. I didn't want to wake up ten years later, at age fifty, and feel like my life was empty, but for more publications, awards, and lab results. I wanted much *more*.

Was that too much to ask? Is each of us destined to be or do just one thing, choose only one path?

And don't we all have many sides? Which side of yourself have you given up on in the course of pursuing your work or family or both at the same time? And given the chance, wouldn't you like to connect with that missing part of yourself—perhaps that creative, fun, exuberant, childlike part of you that rides life like a rodeo cowgirl on a bucking bull? My answer: *"Yes, I would!"*

So in the middle of my life, I began to tackle the seemingly forced separation of my two selves and get happy. Of course, there have been lots of books about what happiness is, and how to make yourself happy. From my reading I gathered that happiness is all about attitude and being able to shift your inner balance of emotions from negative to positive. Happiness also seems to require a certain form of self-permission: as in, give up your attachment to being a stoic victim who is judged only by precisely how productive she is and instead allow yourself to break free, explore, and create. I also learned that being happy is about determination and free will. It's about stepping up and actively claiming your own happiness and not waiting for someone else to mail it to you in a gift basket topped with a big red bow.

But, as an award-winning scientist, I felt like I needed something, well, more substantial, more scientific to really show me the way. Why not apply all that I know about neuroscience to my life? I realized that to be happy I had to use all of my brain—not just the part of my brain that designed great neuroscience experiments. I realized that there were vast parts of my brain that I had stopped using (or used very little) since I started my lab and teaching career at NYU. I had the distinct feeling that these underused parts of my brain were starting to wither away. For example, big parts of the motor areas in my brain were just not being used because I didn't ever move. Parts of my sensory brain, the areas involved in certain (non-science-y) kinds of creativity, and parts of my brain involved in meditation and spirituality were like barren deserts compared to the parts of my brain that designed new experiments, followed the rules, and judged myself at every turn. All of those science-based areas were lush and green with as much life as the Amazon rain forest. I realized I had to get in touch with my own brain—all of it—as a first step toward getting happy. But there was still more.

Despite my deep love and respect for the brain, I also knew that we are more than just our brains—we have a body connected to that brain that allows us to interact with the world. And it wasn't just parts of my brain that were not being stimulated. My entire body was being neglected. I didn't just need to stimulate the barren parts of my brain; what I really needed was to start working my entire body. In essence, I slowly learned that being happy comes down to making sure you not only are using all parts of your brain in a balanced way but are also connecting your brain and your body.

The good news, the amazing news, is this: That when we start activating our brains and start making this mind-body connection, *when we tap into all that our brain does for us and that beautiful and inextricable connection between the body and the brain,* we give ourselves an exceptionable, irreplaceable way to make our brains *work better.* In other words, *we sharpen our thinking and increase our memory capacity.* We learn how to *leverage the good* aspects of our environment (including our bodies) and *protect ourselves* against the bad (stress, negative thoughts, trauma, or addiction).

My own journey began with regular aerobic exercise, with a little yoga thrown in for good measure, after many years of mainly couch potato behavior. Seeing and feeling my body get stronger did something quite magical for me. It gave me a brand-new kind of confidence in my physical being that I had not felt since I was a kid. It made me feel strong and even a little sexy and put me in a fantastic mood that got better the more I worked out. My body was learning new things all the time, and it turns out that my brain loved it! Not only did my mood improve but I found my memory and attention improved as well. I started enjoying life more, my stress levels decreased, and I felt more creative. I even started applying my new passion for physical exercise to my science life, exploring different ways of asking questions and considering new topics on the brain that I had not thought about before. Perhaps the most miraculous thing that happened is that this new confidence, physicality, and great mood started to chip away at the boring, workaholic, controlling, nose-to-the-grindstone "scientist persona" that I had been so lovingly cultivating for so many years. Instead, I felt myself tapping into long-lost passions and embracing joy.

But the secret weapon here to activate your brain and to use the power of the mind-body connection to make ourselves happy, that genie in the bottle, is neuroscience. I came to realize that I was the living example of neuroscience; everything I did with my body was changing my brain, for the *better*! Once this came into focus for me, I knew there was no turning back. I found that when I took the time to develop more of the dimensions of my own identity, I felt more fully myself, more complete. I was 100 percent motivated to make the changes that I needed to make in my life to be happy, to manage my negative thought patterns, stay focused, and follow through on my goals. So what I'm saying is that, from a neuroscience perspective, you *can* use your brain to make yourself happy.

Today I am forty-nine. I am fit, I am happy, I have an active, fun, and surprising social life, and I am as committed to my career as ever before. I travel the globe giving talks and presenting at conferences—to fellow neuroscientists, to medical doctors and their students, to celebrities, and to children of all ages. Because of everyone's fascination with the brain, I am in high demand. I do TED talks and tell Moth stories. I speak to huge crowds of academics. But I have not lost my focus on the regular physical exercise that started me down this transformational path. In fact, I not only teach an undergraduate neuroscience course at NYU that incorporates exercise but I teach a weekly free exercise class open to NYU and the entire New York City community. I walk my talk—literally—every day!

In *Healthy Brain, Happy Life* I want to share with you how I got to this happy place, this life I pined for when I was turning forty. I also want to share with you the science behind this change. From now on, all the neuroscience and brain research you hear about in the news will make sense and feel relevant to you and your life. I am going to offer you advice and insight—based not only on my own experience but also on what all the current and past neuroscience research tells us. This is why I am calling *Healthy Brain, Happy Life* a personal program. It's not a cookie cutter, step-by-step plan; rather, it's a flexible set of accessible advice, tips, and scientific facts that will empower you to change, to grow, to use your very malleable brain as much as possible.

I am also going to share riveting science narratives from my work and the work of legendary scientists in my field that will show you how we've come to understand the brain . . . and what we still don't know.

The chapters contain practical take-aways that get to the central neuroscience concepts that apply to your daily life plus what I call *Brain Hacks*, four-minute shortcuts that act as quick ways to access the brain's power to restore energy, boost mood, and improve thinking. The Brain Hacks make the neuroscience concepts tangible and usable for everyone. For those times when you want a shortcut to your brain, don't have the time or the inclination for exercise, and need a brain boost, use a Brain Hack!

Are you ready to use your brain to jump-start your life? Okay! Let's get started.

HOW A GEEKY GIRL FELL IN LOVE WITH THE BRAIN

The Science of Neuroplasticity and Enrichment

L ong before I wanted to be a scientist, I dreamed of being a Broadway star. My father, an electrical engineer and one of the most diehard Broadway fans you will ever meet, took us to every traveling Broadway production that came to San Francisco, just an hour away from my hometown of Sunnyvale, California. I saw Yul Brynner (when he was about eighty-five) in *The King and I*, Rex Harrison (when he was about ninety-eight) in *My Fair Lady*, and Richard Burton (kind of old, but not ancient) in *Camelot*. I spent my childhood watching Shirley Temple movies and all the classic Hollywood musicals. My dad took my brother and me to see *The Sound of Music* when it was released in the theater each year. We must have seen it twenty times. I fancied myself as a magical blend of Julie Andrews, Shirley Jones, and Shirley Temple, and in my daydreams, I would spontaneously break into song and, in my adorable, impossibly plucky way, save the day and get the guy—all in one fell swoop.

But despite my father's love of all things Broadway, I was clearly expected to do something serious with my life. As a third-generation

Japanese American with a grandfather who had come to the United States in 1910 and founded the largest Japanese-language school on the west coast, my family had high expectations for all of their children. Not that they ever verbalized these high standards—they never had to. It was simply understood that I should work hard at school and pursue a serious career that they could be proud of. And by serious, I knew I had only three choices: I could become a doctor, a lawyer, or something academic—the more impressive sounding the better. I didn't fight these expectations; they made sense to me.

Quite early, in the sixth grade at Ortega Middle School in fact, I began a lifelong pursuit of science. My science teacher that year, Mr. Turner, taught us about the bones of the human body, testing us by having us put one hand into a dark box to identify a bone by touch. I loved it! No squirming for me—I was thrilled by the dare. I became even more excited when I got to do my first pig and frog dissections, and despite the revolting odor, I knew I had to know more. How did all those little organs fit so compactly and beautifully into that little pig body? How did they all work together so seamlessly? If this is what it looked like inside a pig, what might the inside of a human look like? The process of biological dissection captured my imagination from the first moment I got that choking whiff of formaldehyde.

The emerging scientist in me was also fascinated with that most coveted of candy concoctions when I was growing up: Pop Rocks. While other kids in my class were satisfied by the mouth-feel of explosions on their tongues, I wanted to understand *what triggered* these bursts and what wild sensory/chemical experiences you could have in your mouth by combining them with other things, like fizzy seltzer water, hot tea, or ice water. Unfortunately, Mom deemed these experiments a choking hazard and they quickly ended.

My high school math teacher, Mr. Travoli, lovingly guided me through the beauty and logic of A.P. trigonometry. I loved the elegance of math equations, which when done correctly could unlock the keys to a pristine world, balanced on either side of an equal sign. I already had a feeling that understanding math was a key to what I wanted to do (even though I had no idea what that was in high school), and I worked hard to get the best marks in class. In his lilting Italian accent, Mr. Travoli told us over and over again that we advanced-placement students were "the best of the best." I took this as both an encouragement to excel and a solemn responsibility to use my math skills to their fullest potential. I was a serious and earnest kid, on my way to becoming an even more serious teenager.

By this time, the only outlet for my inner Broadway passion was going to the movies. I got my parents to agree to let me see *Saturday Night Fever* on my own by telling them it was a "musical" and conveniently failed to mention the R rating (I was only twelve). They were not pleased when they realized what I had seen. Later, I was obsessed with movies like *Dirty Dancing,* and imagined myself effortlessly stealing the show in Johnny Castle's arms despite the fact that I hadn't taken a single dance lesson since my ballet and tap days in grade school.

By high school, the balance had decidedly shifted. The shining lights of Broadway had dimmed, and I was a steadfast, committed, and driven student, completely at home in a life of total science geekdom. I can see an image of myself in high school: shoulders hunched, serious faced, and carrying a tower of heavy books, as I made my way through the hallways trying not to attract any attention. Yes, I still relived my Broadway fantasies every time I saw one of my favorite musicals on television, but by then those dreams were kept locked in the den at home and studious geek girl had taken over my life. I was entirely immersed in academics, getting straight As and getting into a top college. I had no time left over to even think about my whimsical interests, never mind letting them coexist alongside my devotion to science and math.

I was also painfully shy, never close to being bold enough to date anyone in high school. I was on the tennis team all four years, but how could I not be? My mother was an intense and active amateur tennis player who made sure I played tennis year round and sent me to tennis camp every summer. Tennis was supposed to make me more well rounded, but in reality, what I desperately needed was a camp focused on the topic of how to talk to boys. Well, I never went to *that* camp, and as a consequence, I also didn't go out on a single date or to a single prom through junior high and high school. In other words, if there had been a Miss Wallflower USA contest for nerdy science geeks, I would have blown the competition away.

All those stereotypes about the geeky, dateless science nerd? I proved them true.

FROM BROADWAY STAR TO LAB RAT

Although my science obsession, good grades, and academic drive didn't win me any dates, they did get me somewhere—somewhere good. While I didn't know exactly what kind of science I wanted to pursue, I knew where I wanted to study it. The University of California, Berkeley, just a hop, skip, and jump from Sunnyvale, was my family's alma mater. Yes, I toyed with the idea of moving away to college and even got into Wellesley way on the other side of the country, but I was in love with the beautiful Berkeley campus and quirky-cool vibe of the town and just knew that it was the right school for me. I applied and was successfully admitted, which made me officially the happiest girl in the world that spring. I quickly packed my bags and could not wait to start this new adventure.

It turns out I didn't have to wait long at all to find my academic passion. It came in the form of a freshman honors seminar I took my very first semester at Berkeley called "The Brain and Its

Potential." It was taught by the renowned neuroscientist Professor Marian C. Diamond. There were only about fifteen students in the class, allowing for more direct interaction with the teacher.

I'll never forget the very first day of that class.

First, there was Diamond herself. She looked like a science rock star standing at the front of that classroom, tall, proud, and athletic with a blonde bouffant hairdo that made her look even taller than she was, wearing a crisp white lab coat over a beautiful silk blouse and skirt.

Also, sitting on the table in front of Diamond was a large flowered hat box. After she welcomed us to her class, Diamond threw on a pair of examination gloves, opened the hat box, and slowly and ever so lovingly lifted out an *actual preserved human brain.*

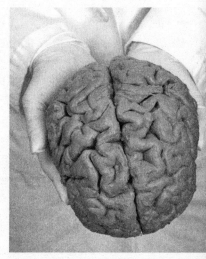

It was the first one I had ever seen in my life, and I was completely mesmerized.

Diamond told us that what she was holding in her hands was the most complex structure known to humankind. It was the structure that defined how we see, feel, taste, smell, and hear the world

The human brain.

around us. It defines our personalities and allows us to go from crying to laughing sometimes in a blink of an eye.

I remember how she held that brain in her hands. This object used to be someone's whole life and being, and she respected that awesome fact in the way she handled that precious piece of tissue.

The brain sported a light tan color that I later learned mainly came from the chemicals used to preserve it. The top part of the brain looked like a compact mass of thick, somewhat unruly tubes.

It had an oblong shape that was slightly wider on one end than the other. When she turned the brain to the side, I could see more of the complexity of the structure, with the front side of the brain shorter than the back end. The divided and paired structure of the brain was obvious at first glance—the right and left sides of the brain were each separated into different parts, or lobes.

THE BRAIN AND ALL ITS PARTS

Neuroscientists used to think of the different parts of the brain as housing certain functions. We know now that that's only partially true. While specific areas of the brain do have specific functions (see the following list), it's important to keep in mind that all parts of the brain are connected, like a vast and intricate network.

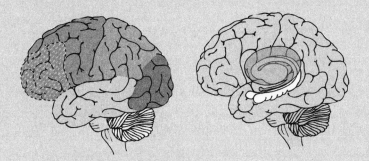

- **Frontal lobe:** This front section of the brain houses the all-important prefrontal cortex (making up the front part of the frontal lobe), understood to be the so-called seat of personality and integral to planning and attention, working memory, decision making, and managing social behavior. The primary motor cortex, the area responsible for allowing us to move our bodies, forms the most posterior (toward the back) boundary of the frontal lobe.
- **Parietal lobe:** This lobe is important for visual–spatial functions and works with the frontal lobe to help make decisions. The part of the cortex responsible for allowing us to feel sensations from

our bodies (known as the primary touch cortex) is located at the most anterior (toward the front) part of the parietal lobe.

- **Occipital lobe:** This is the part of the brain that allows us to see.
- **Temporal lobe:** This is the part of the brain involved in hearing, vision, and memory.
- **Hippocampus:** Located deep inside the temporal lobe, this area is crucial for the formation of long-term memories; it's also involved in aspects of mood and imagination.
- **Amygdala:** This structure, which is critical for the processing of and response to emotions such as fear, anger, and attraction, is also located deep inside the temporal lobe right in front of the hippocampus.
- **Striatum:** This area, which is seen best from a cut down the middle of the brain, is involved in motor function and plays an important role in how we form habits (and why they are so hard to break!); it's also integral to the reward system and how addictions develop.

Like the best teachers do, Diamond then made what initially seemed incomprehensibly complex totally understandable. She told us that this big complex mass of tissue was really made up of only two kinds of cells: neurons and glia. Neurons are the workhorses of the brain and each contains a cell body, which is the control center of the neuron; input structures called dendrites, which look like big tree branches, that receive information coming into the cell body; and a thin output structure called the axon, which can also have lots of branches.

What makes neurons unique from any other cell in the body is that they are able to communicate via brief bursts of electrical activity, called action potentials, or spikes. That cross talk between the axon of one neuron and the dendrite of the next one takes place at a special communication point between the two called a synapse. It's the brain's electrical "chatter," or axon-to-dendrite communication, that is the basis for all the brain does.

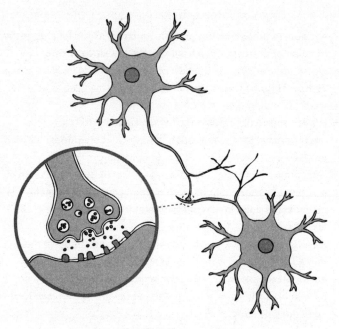

Neurons and their connections.

What about the glia cells? *Glia* means "glue," and the cells were so named because scientists in the nineteenth century mistakenly thought these cells had something to do with holding the brain together. While it's true that some of the glia cells do serve a scaffolding function in the brain, we now know that they actually serve a wide range of different support functions for neurons. Glia cells supply nutrients and oxygen to the neurons; they form a special coating on the neurons called myelin, which is required for normal synaptic transmission; and they attack germs and serve as the brain's cleanup crew, removing the debris from dead neurons. Exciting new evidence suggests that glia cells may even be playing an important role in certain cognitive functions including memory. Many believe there are ten to fifty times more glia in the brain than neurons, but this often-repeated statistic is being challenged by new studies suggesting that the ratio is closer to one to one.

Diamond then explained that if we had a big bucket of neurons and another big bucket of glia, we, at least in theory, would be able to build a brain. But the big puzzle is figuring out exactly how we put those neurons and glia together to work as beautifully and elegantly—as perfectly and imperfectly, as correctly and incorrectly—as a real brain. I learned that day that figuring out those connections and the general question of how a brain is put together, otherwise known as the study of neuroanatomy, was Diamond's specialty.

But what truly captivated the nascent scientist in me that first day of class was her description of brain plasticity. This does not mean your brain is made of plastic, but rather it refers to the idea that the brain has an essential ability to change (like a piece of malleable plastic) as a result of experience. And by *change* she meant the brain could grow new connections within itself. I still remember her giving us the analogy that if you study really hard your brain may ache because of all the axons and dendrites growing and straining to make new connections.

In fact, Diamond (as one of the very few women in science at the time) had been responsible for the now classic research starting in the early 1960s on exactly how plastic, or malleable, our brains really are. At that time, it was known that brains could change and grow extensively from infancy to adulthood, but it was believed that once we hit adulthood, our brains were set in stone, with no ability to grow or change.

Diamond and her colleagues at Berkeley challenged this notion in a very big way. In their now famous study, they asked what would happen to the brains of adult rats if you housed them in what she called "enriched environments." This meant letting them live in a sort of Disney World for rodents, with lots of colorful toys to play with, lots of space to run around in, and lots of other rats to engage with. The researchers were looking to topple the idea that the adult human brain was fixed—that is, that it was not capable of change. In order to answer their question, Diamond and her team changed the physical environment that the rats lived in and asked whether there was any effect on the physical structure of the brain. If there was evidence of change in the rats, then that meant under certain conditions, human brains might also be able to grow, adapt, or change.

What were the results of housing rats in Disney World? Compared to rats living in what the researchers called impoverished environments, with no toys and only a few other rats to play with, the rats living in Disney World actually had brains that were physically *larger* than the impoverished rats. Diamond showed that in the enriched environment, dendritic branches (those input structures of the neurons that look like tree branches) actually *grow* and *expand,* allowing the cells to receive and process larger amounts of information. In fact, she showed that not only were there more dendritic branches but more synaptic connections, more blood vessels in the brain (that means better access to oxygen and nutrients), and higher levels of good brain chemicals like the neurotransmitter acetylcholine and particular growth factors.

Diamond explained that these differences in brain size were a direct reflection of the nature of the rats' environments. In other words, the size and function of a brain—rat or human—is highly sensitive and reactive to all aspects of any given environment— physical, psychological, emotional, and cognitive. This constant interaction between the brain and the environment, combined with the brain's ability to respond by changing its anatomical structure and physiology, is what neuroscientists mean by the term *brain plasticity.* Stimulate the brain with new things to do or new individuals to interact with and it reacts by creating *new connections* that cause it to actually *expand in size.* But deprive your brain of new stimulation or bore it with doing the same thing day after day after day, and the connections will *wither away* and your brain will actually *shrink.*

In other words, your brain is constantly responding to the way you interact with the world. The more diverse and complicated your interactions, the more neural connections your brain will make. The less enriched your environment and experience, the fewer neural connections your brain will make. There was nothing particularly special about the rats raised in Disney World; in fact all the rats in the study had the same capacity for reaction to stimuli. Do you play the piano? Then the part of your brain that represents the motor functions of your hands has changed relative to people who don't play the piano. Do you paint? Play tennis? Bowl? All of these things we also know change your brain. We now understand that even the everyday things that we learn—the name of the guy that takes our order at Starbucks or the name of the newest movie we want to see—are all examples of the brain learning, which in turn causes the brain to make micro changes in its structure.

It was almost too much fascinating information to take in for the first day of class. But one thing was for sure. The first day of "The Brain and Its Potential" class changed my life. I walked in a curious enthusiastic freshman wanting to soak it all in, and I walked out a curious enthusiastic freshman with newfound purpose and meaning. I knew after that day in class what I wanted to do with my life. I wanted to study that lumpy mass of tissue and discover some of the secrets to understanding what it is to be human. I wanted to be a neuroscientist.

Over the next four years, I took many more classes with Diamond, including her wildly popular gross human anatomy class and her more advanced neuroanatomy class. You might not realize how much passion, enthusiasm, and clarity (plus a little magic) it takes to make an anatomy class really interesting. A course in gross human anatomy is a practice of committing to memory every single detail of your body—from your bones to your muscles (including the specific locations where bones attach) as well as every single internal organ and how each of them is hooked up to another. There are more than seventy-five hundred parts to the human body! As you can imagine, memorizing every single one is an enormous task. If a professor simply presented all this anatomical information in a flat, listlike way, the class would be akin to reading this year's new income tax regulations—dry as dirt. But Diamond revealed the human body to us as if we were on a grand adventure in an exciting new universe, both familiar and strange. She also made everything personal, telling us that learning about the anatomy of our bodies was going to teach us about who we were as people. If we were going to keep our anatomy and our brain for the rest of our lives, wouldn't it make sense to know what we were working with?

Diamond was a master at mixing information about the origin of an anatomical term or some lesser known anatomical fact with more basic lessons, therefore making every last piece of information seem relevant.

For example, she asked us:

The word *uterus* in Latin means "hysteria." Do you agree
with this?
Do you know what the largest organ in the body is? It's your
skin! Take care of it!
Isn't the psychology of hair and hairstyles fascinating? We
could have a whole course just on that!

With every comment and through every lecture, she made anatomy personal and come to life. I remember in the middle of the semester I took gross human anatomy, I happened to go see the Alvin Ailey dance troupe perform for the very first time at Zellerbach Hall on the Berkeley campus. That was the first time I saw their famous piece *Revelations*. Not only was I mesmerized by the dancing that

night but, because we had just been going over the origins and in-
sertions of all the muscles of the leg, I could now appreciate all those
movements on a whole different anatomical level. To me there was
no better example of the beauty of the human body than the shapes
and movements that I was seeing on stage.

Diamond was truly an inspiration. It was so clear she loved and
appreciated the topics she was teaching, and she genuinely wanted us
to love and appreciate this breadth of information in the same way.
She didn't only care about the subject matter but cared deeply for us
students as well. She was more than available to answer questions,
and just to be sure she got to know at least some of us in her class of at
least 150 students, she would randomly pick names of those enrolled
out of a hat and take two of them out to lunch just to chat over a meal.
When I was taking her class all her students also had an open invitation to come out to the tennis courts on the north side of campus to
play an early morning set of doubles with her anytime. This sounds
like the perfect invitation for the tennis-playing neuroscience geek
from Sunnyvale, right? Well, I have to admit that I let my shyness get
the better of me, and I never gathered the courage to go play tennis
with her in all the years I went to Berkeley—to this day it's one of the
biggest regrets on my list of should-haves from my college years.

Some of her teaching magic started to rub off on me, even back
then. I remember an afternoon practical session where we had a
whole bunch of organs spread out at different stations in the room
that we were supposed to examine and learn about. I was particularly
intrigued with the dense, multilobed liver and the little nub of a bile
duct hanging off the bottom. I remember figuring out all the parts of
the liver we had learned in class and another student coming by and
asking me what we were supposed to be seeing here. I explained to
him everything that I had discovered on this example liver, and he
seemed to get it quickly. I ended up spending the next thirty minutes
presiding over that liver and explaining to any and all students that

The author with Dr. Marian Diamond on the day Suzuki graduated from Berkeley.

came by all the key features of the organ. That day I became a liver anatomy expert. I think that was the day I also became a teacher. And I learned a valuable lesson that was going to serve me well for the rest of a career: The best way to learn something deeply is to teach others about it. I use that principle to this day.

I was certainly not the only one to love Diamond's gross human anatomy class. On the last day of class, several other students came to class with flowers and literally threw them at her feet! I was there cheering and shouting right along with them, celebrating the end of this great course, my only regret being that I hadn't thought to bring flowers to throw.

CHECK OUT MY ROCK-STAR PROFESSOR!

The great thing about our digital age is that you now can experience some of Diamond's classes yourself. Just search for "Marian Diamond" on YouTube. Check her out!

WHAT WE KNOW ABOUT THE BRAINS OF CAB DRIVERS

We have come a long way in our understanding of brain plasticity since Marian Diamond's early enriched environment studies in rodents. Now there is lots of evidence of brain plasticity, including in humans. One of my very favorite examples of adult human brain plasticity was done by my colleague Eleanor Maguire, at University College London. Maguire didn't send her human subjects to live in Disney World for a year. Instead she studied a group of people who had meticulously learned a very specific and extensive body of knowledge about their home turf. Namely, she studied London taxicab drivers. You see, London cabbies have the daunting task of learning to navigate the more than twenty-five thousand streets in central London as well as the locations of thousands of landmarks and other places of interest. The extensive training period that is required to learn all this spatial information is called "Acquiring the Knowledge" and typically takes between three and four years of study. If you have ever been to London and seen people riding around on scooters with maps splayed out on the handle bars, those are the aspiring London taxicab drivers learning these skills!

Only a fraction of the aspiring cab drivers actually pass the stringent exams, called, very dauntingly, "Appearances," but those who do pass demonstrate an impressive and extensive spatial and navigational knowledge of London. What an interesting group of people (and brains) to study!

In the study of these London cab drivers, Maguire's group focused on the size of a brain structure that I will be discussing a lot in the upcoming chapters: the hippocampus. This is a long seahorse-shaped structure deep in the brain's temporal lobe (*hippocampus* means "seahorse" in Latin) critically involved in long-term memory function, including spatial learning and memory. More specifically, because Maguire and her colleagues had localized spatial memory function to the back, or posterior, part of the hippocampus, they wondered if that part of the brain structure might be larger than the anterior (front) part in cab drivers when compared to control subjects who were matched for age and education. In fact, that's exactly what they found.

Maguire's research and other studies that have compared the brains of experts (such as musicians, dancers, and people of particular political affiliations) to nonexperts have all been used as examples of brain plasticity in humans. While plasticity is one possible interpretation of the data, another possibility is that people who succeed as London taxicab drivers have larger posterior hippocampi to start with. In other words, it could be that only people with naturally big posterior hippocampi have the superior spatial navigation ability required to succeed as a London cabbie. If this were true, it would not be a case of brain plasticity at all.

So, how can we differentiate between these possibilities? What would need to happen to test the idea that the experience of learning to be a London taxicab driver changes the brain would be to actually follow a group of people who started Acquiring the Knowledge and then compare the brains of those who eventually passed the test with those who did not. And that's exactly what Maguire and her team did. This kind of study is much more powerful because you can clearly identify any brain changes as a function of taxicab training. What the researchers found is that before training started, all the wide-eyed and bushy-tailed London taxicab driver wannabes had the same size hippocampi. The scientists then reexamined the cab drivers after they had completed the training period and after they knew who passed and who failed. They found that the wannabe cabbies who passed now had significantly larger posterior hippocampi than they did before they started their training. *Ta-da!* Brain plasticity in the flesh! This group's posterior hippocampi were also larger than those of the subjects who hadn't passed. In other words, this experiment showed that successful training to pass the Appearances exam did indeed enlarge the hippocampus, and the trainees who had not retained enough information showed far less of an increase in size.

This is just one example of the everyday, beautiful plasticity of our brains. Everything we do and for how long and intensely we do it affects our brains. Become an expert bird watcher, and your brain's visual system changes to be able to recognize all those tiny little birds. Dance tango all the time, and your motor system shifts to accommodate all those precise kicks and flicks you are doing with your feet. The life lesson I learned all those years ago in Diamond's classroom was that I shape my brain every day and so do you.

MY OWN DOORMAN EXPERIMENT

London is not the only city where its municipal workers have special skills. In New York, it's doormen. Think about all those faces they have to recognize and differentiate from strangers if they work in a thirty- or forty-story high-rise! Here is a thought experiment I would love to do with New York City doormen if the opportunity ever arose. I would examine the doormen's brain areas known to be important for face recognition and compare the size of that area to those of other city workers who don't have to remember lots of faces (let's say subway conductors). Where exactly is the face recognition area in the brain? At the bottom of the temporal lobe is a unique area known as the fusiform face area, which specializes in helping us recognize faces. When this region is damaged, people cannot distinguish facial features, a condition known as prosopagnosia. The actor Brad Pitt, the famous portrait painter and photographer Chuck Close, and the Harvard professor and author of *Multiple Intelligences*

Howard Gardner are a few famous people with prosopagnosia. Because they cannot recognize people by their faces, they rely on other features such as voice, hair, gait, and clothing. But in doormen, who develop and hone the skill of quickly recognizing sometimes hundreds of faces, I predict that this fusiform area will be significantly larger than that in the subway conductors. Maybe someday I'll get to do this experiment.

MY OWN PERSONAL ENRICHED ENVIRONMENT: ADVENTURES IN BORDEAUX

My life in college was firmly focused on doing well in my classes, though I did date a couple of guys (somewhat awkwardly) during my first two years at Berkeley. Despite my general shyness as a young woman, the truth is that I have always had an adventurous spirit and I was itching to see the world and travel abroad. U.C. Berkley had the perfect study abroad program for me, and I signed up in my junior year. I learned that if I went to particular campuses abroad, I could even take science classes that counted toward my major of physiology and anatomy, so I would not lose any credits. The only country I would even consider visiting was France. I had been enchanted with the French language ever since I started learning it in junior high. My choices for campuses were either Bordeaux or Marseille. In other words, wine or fish soup. The choice was clear—I went for the wine! Little did I know when I signed up for my junior year abroad adventure that France, with its unique culture, beautiful language, strong traditions, glorious foods and wines, stylish clothes, amazing museums, excellent educational system, and brilliant residents (especially the men) was going to serve as my own personal enriched environment for the next twelve months.

IS THERE A CRITICAL PERIOD FOR LEARNING A LANGUAGE?

Everyone agrees that there is a very special time, called the critical period, during about the first six months of life when the brain is particularly good at learning languages. Wonderful work from Professor Patricia Kuhl at the University of Washington has shown that babies' brains can soak up and learn not just one language but multiple languages during this time.

But what if you start learning a new language a little later in life? Like most people of my generation, I began learning a second language (in my case French) at the ripe old age of twelve when I got to middle school. What part of my brain helped me learn this second language? It turns out that the brain does rely on many of the same areas as are used when learning to speak your native tongue. However, you also seem to recruit additional brain areas to help you with a second late-learned language. These additional areas are situated toward the bottom part of the frontal lobe on the left side, called the inferior frontal gyrus. You also use the left parietal lobe. Another study showed that people (like me) who learned language later in life actually had a thicker cortex in the left interior frontal gyrus and a thinner cortex in the right inferior frontal gyrus.

Learning a second language at twelve years or later provides yet another example of brain plasticity. The brain, when prodded to make connections, will indeed do so. It might take longer and be more difficult, but it's possible!

I loved the year I spent in France because it completely immersed me in a totally foreign and exotic culture that in 1985 was far less infiltrated by American cultural icons like McDonald's, Costco, and reruns of *Friends* than you see in France today. That year abroad also brought me one of the most romantic experiences of my life.

It all started with my request to live with a family in Bordeaux who had a piano that I could play. I had played the piano from the time I was about seven until I was a senior in high school, and I still played casually (so as not to completely lose my classical repertoire) while I was studying at Berkeley.

Monsieur and Madame Beauville were a lovely couple whose home had a few extra bedrooms upstairs, one of which housed a piano. Soon after I arrived, Madame Beauville asked me to make sure I was at home one particular afternoon at a particular time because she had hired a piano tuner to come. I happily agreed and waited for the little old man with white hair to come walking up the stairs to my bedroom to tune the piano. But to my surprise, it was not grand-père who made his way up the stairs to my bedroom, but a young, hot French guy named François. François set about tuning my piano and chatting with me in French, of course. Before that day I never thought I was particularly good at flirting. But that day I discovered I *was* good at it, and I could even do it in French! In that hour, I not only got a perfectly tuned piano for my bedroom but I snagged a card with the address of a sheet music shop where François worked part time and an invitation to come by and say bonjour anytime.

Of course, I somehow found time in my busy schedule of lectures, coffee, and croissants to visit him in the music shop right around dinnertime, and he invited me for a bite to eat. After just a few more dates that began after his shift in the music shop, we were an item, and I suddenly had myself a very sweet and musically inclined French boyfriend.

How had I come out so far out of my shell? I had no idea, but I see now the enormous amount of brain plasticity occurring that year. This was even better than living in Disney World. Everything was so different—not only was I speaking French all the time and taking all of my classes in French but I truly felt like a different person while speaking the language. Suddenly, I was no longer the geeky wallflower with nary a date in sight. Instead, in France I was considered extremely exotic because I was an Asian woman from California who didn't speak Japanese but instead spoke fluent American. When I was growing up in northern California, Asian American women were a dime a dozen; now I got to be exotic for the very first time in my life. That was huge for me. Not only that, but—I don't know if you are aware—the French kiss each other all the time. It's a rule. You have to kiss; it's frowned upon if you don't. Finally! An excuse to kiss everyone for the girl from the family that didn't hug or kiss much at all—I was in seventh heaven.

And the more I learned, the happier I became.

In France, all this kissing made me step out of my comfort zone and become a lot looser and a *lot* more affectionate. I now realize that making these kinds of changes literally expanded who I was: As I changed my behaviors and experienced new sensations, my brain made adjustments to this new information and stimuli.

Aside from François, my French became fluent because I was also taking some serious science classes—not with American students mind you, but with all the other French students. That meant all the lectures were in French and, most terrifying for me, the oral and written exams were all in French. I was not that worried about the written exams because most of the science words are the same as or similar to the English words. But I had never in my academic career taken an oral exam. Much less in my second language. I was totally scared.

One of my clearest memories from this time was while I was responding to the questions a professor posed during an oral exam. I was very nervous and suddenly lost all ability to speak with a proper French accent. The words and the grammar coming out of my mouth were all French, but the sounds were pure American. I could hear myself speaking French with an incredibly strong American accent—*quelle horreur!* Good thing I was graded on content and not verbal presentation. I ended up acing all my classes. Clearly the geeky bookworm was still present somewhere in my new French incarnation.

This experience in France also gave me another unexpected and what turned out to be lifelong gift. It was in France that I became fascinated with the study of memory, another form of brain plasticity. I had the great good luck to take a course at the University of Bordeaux called "La Neuropsychology de la Memoire" (neuropsychology of memory). This course was taught by a very well-respected neuroscientist, Robert Jaffard, who not only ran an active research lab but was a wonderfully clear and engaging lecturer. I had no idea there was a strong neuroscience group at the University of Bordeaux when I chose it, but what a lucky coincidence. Jaffard was the first to teach me about the history of the study of memory and the raging debates of the day involving two researchers at the University of California, San Diego, named Stuart Zola-Morgan and Larry Squire and one researcher at the National Institutes of Health (NIH) named Mort Mishkin. Little did I know at the time that in the next ten years I would work with all three of these scientists either as a graduate student at U.C. San Diego or as a post-doc at NIH. Most important, Jaffard took student volunteers in his lab, and I happily began testing little black mice on memory tasks in my spare time, giving me my very first real taste of laboratory research. I loved it in the lab, and this, along with the wonderful background in neuroanatomy I had from Diamond (I also worked my entire last year at Berkeley in Diamond's lab), made it easy for me to decide that I wanted to apply to graduate school as soon as I finished my undergraduate degree.

In between studying for classes and working in the lab, there was François. It turns out that he not only tuned pianos but played the piano and had a near obsessive fascination with the harmonies of the Beach Boys. So I had found a French guy with California in his heart. He had tapes of all the Beach Boys albums and I would often find him in his living room listening intently to them through his headphones as he tried to painstakingly transcribe all the complex chords they used to create their sound. He was doing this with such glee and concentration that I hated to interrupt his sessions. I too was a big Beach Boys fan, but I had never fully appreciated the complexities of their harmonies before François. I thought the Beach Boys were just fun and easy to dance to, but François, with his trained musical ear showed me his favorite chords and riffs in that music I knew so well, and opened it up in a very different way for me.

One of the many things we enjoyed doing together was playing piano duets. At first, François had only one piano in his apartment, but because he worked at the biggest piano store in town, he eventually borrowed a second piano so that we could practice and play our duets in his apartment, where I was spending increasing amounts of my time. And because I loved playing classical music, we played classical duets—Bach, to be precise.

But the really fun part was when we went to the piano shop at night after it had closed. There in the empty store we performed our duets on the big beautiful eight-foot concert grand pianos that were used for performances in the local theaters. I always played the Bösendorfer (I loved the sound of those low notes), and he played the Steinway. We played as loud and long as we wanted, and the beautiful tones of these pianos (expertly tuned by François himself) made even the mistakes sound good. I consider these evenings as some of the loveliest times I spent with François.

In addition to playing classical music together, we listened to a lot of it. One of my favorites was the Bach solo cello suites. I listened to François's record of Yo-Yo Ma playing these pieces over and over again. It turns out that François noticed how much I loved them, and that Christmas I received the most precious gift that I had ever received before or since: a cello.

I was flabbergasted.

For someone who had dated only a little bit in her first two years of college, I was getting a crash course on romance from François that I didn't want to end. I decided that the myth was absolutely true: The French *are* the most romantic people in the world!

THIS IS YOUR BRAIN ON MUSIC!

Do you ever wonder what happens in your brain when you hear that piece of music you can listen to over and over and over again? The one that may even give you chills just listening? Robert Zatorre and his colleagues at the Montreal Neurological Institute showed that when people listened to music that gave them a strong emotional and physiological response (the Beach Boys for François, and Bach for me), the brain showed significant changes in the areas involved in reward, motivation, emotion, and arousal: the amygdala, orbitofrontal cortex (the bottom part of the prefrontal cortex), ventral medial prefrontal cortex, ventral striatum, and midbrain were all activated. So as François and I delved into playing and listening to music together, we were also activating the reward and motivation centers in our brains (see Chapter 8). No wonder I loved France so much!

So my French-enriched environment gave me a new language, a new persona, romance, adventure, and—of course, I have to add to the list—great food and wine. It was during this time and with François that I also really developed my love of French cuisine. My parents, and in fact my whole family, are great lovers of food and any big celebration—whether it be a graduation or a recital—has always been celebrated at a wonderful restaurant. But in France, my food experiences stepped up to a whole new, more sophisticated level. While I was a poor college student, you could still eat (and drink) like a king in Bordeaux, especially if you had a native son like François as a guide. Yes, I worked and studied like the science geek that I was, but I ate, drank, and spent leisure time playing the piano like a sexy flirtatious exotic French woman in love. Look at me! The world-class wallflower from Sunnyvale had a fantastic French boyfriend and a rich social, food, and cultural life. It was easy to do in such an enriching, stimulating environment.

FOOD, WINE, AND BUILDING NEW BRAIN CELLS

Living in France, it was not hard to eat a lot of delicious, flavorful French food and drink many delectable bottles of wine. Indeed, I tasted, sipped, and enjoyed wines of all kinds from all over France— from Burgundy, the Loire Valley, Provence, and Bordeaux. White, red, rose, and of course Champagne. All of these new tastes were literally turning on my brain. It turns out that experiments in rodents have shown that enriching your olfactory/gustatory environment does have a significant effect on the brain.

Studies show that once we grow into adulthood, there are only two brain areas where neurogenesis (the birth of new neurons) can occur. The first brain area is the hippocampus, which is crucial for long-term memory and mood (more on these two features in upcoming chapters) and the second is the olfactory bulb, the brain area that is responsible for our sense of smell and therefore also contributes to our sense of taste. Studies show that if you enrich the olfactory environment of rats by giving them a nice big range of smells, you can enhance neurogenesis in the olfactory bulb and that the brain actually increases in size because of these new neurons. This suggests that my French adventure was not only teaching me a lot of about food and wine appreciation, but might actually have been enhancing the size of my olfactory bulb. While changes in size of the olfactory bulb in people with enhanced olfactory experiences have never been explicitly studied, it would be fascinating to examine this potential form of human brain plasticity. I feel a new brain plasticity experiment with sommeliers coming on!

In short, I loved France. I loved my life with François. But as the year passed by I knew that soon I would have to face the reality that I was due back at U.C. Berkeley to start my critical senior year and begin the next phase of my life. This was a difficult time for me because from a very early age I have always had a hard time letting go. I was the kid who worked herself into a tizzy and cried at the end of the summer because I didn't want it to end and to go back to school. And I *loved* school. I just didn't like endings. I think it was the fear that if something wonderful like summer vacation ended, I would never get it back. I don't know where this fear came from—maybe I had a toy taken away from me as a child—I can't say for sure. But what I do know is that I had that terrible feeling of impending sadness in the spring of 1986 when my year in France was coming to an end.

In fact, I seriously thought about staying in France to finish my college career and do graduate work there. That would work, right? I was already working in a lab. A wise French scientist working in the Jaffard lab, to whom I will always be grateful, convinced me that I would be much better off going to an excellent graduate program in the United States. He was right, but that was not the answer I wanted to hear. My parents, who were none too happy about my involvement with a piano tuner/musician with no degree or higher education of any kind, wanted me back home and attending classes at U.C. Berkeley immediately.

But I didn't want this magical year or my relationship with François to come to an end. How could I? Give up being the exotic English-speaking Asian girl with the hot French boyfriend for my old single life of science geekdom in the States? What could be worse?

I knew I was at a crucial point in my life. I knew that yes, I did have to return. There was no real possibility of staying in France after my year was up. I knew deep down that I not only *had* to finish my degree at Berkeley but I really *wanted* to finish my degree at Berkeley. But François could come with me, right? We could be together in the United States, and then we would decide what came next. Both of us held on to that dream for several months after I returned to school in California and started taking classes again. We were like a French and Japanese American version of Romeo and Juliet, with my parents playing the part of the disapproving family perfectly. Actually, my parents had enough disapproval between them to play the roles of both the Capulets and the Montagues. Every day, I wrote François long letters in French, telling him about everything that was so different in the States and all the things I missed about living in Bordeaux. We both wanted to keep our relationship alive, me with the hot French musician and he with the exotic Asian American girl who loved science.

That dream lasted for several months, until one day reality knocked on my door and walked right in. Specifically, the reality of applying and attending graduate school arrived. I suddenly realized that it was unlikely that François could make a living tuning pianos in the United States, especially since he didn't speak English. And most difficult for me to admit, I knew deep down that while I so enjoyed going out with him for the year that we were together, he was probably not the lifelong partner for me. Besides, what did I know at the ripe old age of twenty-one? He was the first serious boyfriend I had ever had.

To this day, my last phone conversation with François is etched in my memory with great detail. I remember where I was sitting in my little studio apartment in Berkeley; I remember how I was sitting and holding the phone. Mostly I remember the pain, guilt, and discomfort I felt during that conversation, as if it happened yesterday. I did a terrible job breaking up with him and I knew it, but at that time I didn't know any other way. I should have been more loving and understanding, and I should have explained the situation and my logic more clearly. Instead I felt pressured to get on with my life, and I was rude and abrupt with him. I know why I remember that call in so much detail. Emotion, either very negative, like the one that I was experiencing that day, or very positive, helps strengthen memories. One brain structure in particular, called the amygdala, which sits in the temporal lobe just in front of the hippocampus, is critical in the formation of strong memories from strong emotions. My amygdala was working overtime that day (you'll learn much more about why we remember emotional events in the next chapter).

That day, I chose science over François. It was a hard decision, and it took me months to recover. But I know now that it was a choice that shaped the rest of my life.

THE STAR OF OUR EVOLUTIONARY BRAIN

The prefrontal cortex (PFC), situated just behind the forehead, was the last part of the brain to evolve, and scientists agree that it sets humans apart from most other animals. The PFC is essential for some of our highest-order cognitive abilities, including working memory (defined as the memory we use to keep things in mind, also referred to scratchpad memory), decision making and planning, and flexible thinking. In essence, this is command central for all of our executive functions, which play a role in so much of what we do and how we think. You will see how the PFC has a role in applying new concepts to other learning situations, managing our stress response, and supervising our reward system. Keep an eye out for the powerful PFC!

TAKE-AWAYS: BRAIN PLASTICITY

- The brain is made up of only two kinds of cells: neurons (brain cells) and glia (supporting cells).
- Brain plasticity is the ability of the brain to change in response to the environment. Raising rats in enriched environments results in a thicker cortex, more blood vessels, and higher levels of certain neurotransmitters and growth factors.
- Training as a London taxicab driver results in brain plasticity. Cab driver recruits who studied and passed the difficult qualifying examination had larger posterior hippocampi, a structure known to be involved in spatial memory, than those who did not pass the exam.
- Areas of the brain recruited when you lean a second language include the inferior frontal gyrus on the left side and parts of the parietal lobe on the left side. Language in general is controlled by the left side of the brain.
- Music activates the parts of the brain involved in reward, motivation, emotion, and arousal, which include the amygdala, orbitofrontal cortex, ventral medial prefrontal cortex, ventral striatum, and midbrain.
- The prefrontal cortex is the command center of the evolved human brain and supervises all executive functions.
- Enriching your olfactory environment with lots of different smells stimulates the growth of new brain cells in the olfactory bulb, a key part of the brain responsible for our sense of smell.

BRAIN HACKS: HOW DO I ENRICH MY BRAIN?

You may not have time to go live in Disney World or France for the next few months, but the great news is that you can start to enrich your brain with these Brain Hacks, most of which take no more than four minutes per day.

- **Motor cortex Brain Hack:** Go online and teach yourself a new dance move from the *So You Think You Can Dance* website and then practice it for four minutes to your favorite music.
- **Taste cortex Brain Hack:** Try a cuisine that you have never tried before: Laotian, African, Croatian, and Turkish come to mind. Be adventurous! And here is another taste cortex Brain Hack for extra credit: Try eating a meal in complete darkness and see how the lack of visual input affects your sense of taste. It should change your experience of the meal and allow a pure taste sensation to come through.
- **Cognitive Brain Hack:** There are so many fun possibilities to enrich your brain. Here are just a few: Watch a TED talk on a topic you know nothing about. Listen to a story from the *Moth Radio Hour*, a storytelling program with a wide range of topics. Listen to a popular podcast that you have never listened to before. Read a story from the section of the newspaper that you never read—for me it would be finance or sports.
- **Visual cortex Brain Hack:** The next time you go to a museum, pick a piece of artwork that you are not familiar with and just sit quietly and get lost visually in it for at least four minutes. In reality, it could take hours to really explore a new piece fully; you can get a great start, though, in just four minutes. A hack for this hack is to simply find a new piece of art online and explore it visually on your computer. Both will stimulate your visual cortex.
- **Auditory cortex Brain Hack:** Go to iTunes, the YouTube music channel, Pandora, Spotify, or whatever music site you like and listen to a really popular song from a genre of music you never listen to or in a different language. Try to understand why it might be number one for that genre.
- **Olfactory Brain Hack:** The main difference between regular sommeliers (who can differentiate many different scents and describe them so precisely) and you or me is one thing: practice. Take just a few minutes to sit and smell your most odorous meal of the day. It might be breakfast with a rich, aromatic cup of coffee and the deep comforting smell of toast fresh from the

toaster, or it might be your dinner of chicken tikka masala from your favorite Indian restaurant. Before digging in, take a few minutes to smell the food and try to really notice the different aromas and try to describe them. This will start to tune you in more to your olfactory senses.

SOLVING THE MYSTERIES OF MEMORY

How Memories Are Formed and Retained

After spending my senior year back at U.C. Berkeley in Diamond's lab, I completed my senior thesis, which focused on examining the brain sizes of rat babies after placing rat mothers in enriched environments. I then graduated with high honors and I knew I wanted to go to graduate school and learn how to become a neuroscientist myself. My time in the Jaffard lab in Bordeaux had also piqued my interest in the brain basis of memory. After all, memory is one of the most common categories of brain plasticity. We know that every single time we learn something new, something in our brain changes. But for me, at the beginning of graduate school, the question was *how* does the brain change. I was also interested in another question: Could we find a way to visualize what was happening the moment something was learned?

The many facets of memory intrigued me. In an intuitive way, I understood that when the brain learns something new, it must change. But where was this happening? What are the challenges to learning something new? And what did learning have to do with memory? My hunch was that all these questions had to do with how memories are structured and formed in the brain. When I was accepted into the graduate program for neuroscience at U.C. San Diego, I wanted to discover everything there was to know about memory; it turns out I was about to become involved in one of the most dramatic and far-reaching areas of neuroscience research.

A SEISMIC SHIFT IN OUR UNDERSTANDING OF MEMORY AND THE BRAIN

U.C. San Diego had a top-notch neuroscience faculty, including Larry Squire and Stuart Zola-Morgan, two neuroscientists whom I had first learned about in Jaffard's course on memory. I didn't know it the day I accepted the offer, but I was soon going to be in the eye of the firestorm over memory function that Jaffard had described in class.

The late 1980s was an electric time to be studying memory. A gigantic memory mystery had emerged in the field centering on the question of a specific brain region that was really critical for memory. This mystery had actually started thirty years earlier, with the most famous amnesic patient ever studied, a man known as H.M.

At the heart of this groundbreaking discovery in the 1950s was a neuroscientist named Brenda Milner, a Brit who got her degree at Cambridge University and was working at McGill University in Montreal, Canada. Milner, an assistant professor at the time, had been working with the eminent neurosurgeon Wilder Penfield, who specialized in brain surgery for serious cases of epilepsy that did not respond to drug treatment. This surgical intervention involved removing the hippocampus and amygdala from the side of the brain where the specialists thought the seizures started. Milner was testing Penfield's epilepsy patients before and after their brain surgery to see if removing the hippocampus and amygdala had any adverse effect on their brain function. She found mild memory impairment for spatial information if the right hippocampus was removed and mild verbal memory impairments if the left hippocampus was removed, but these deficits were considered acceptable given that the surgeries greatly reduced or eliminated the devastating epileptic seizures that these patients had been suffering for years before the operation.

And then the team was absolutely flabbergasted when Penfield treated two new patients with the same brain surgery and got completely different results: After surgery, these patients showed profound and devastating memory deficits. Penfield and his colleagues had done more than a hundred similar operations with only mild memory impairments. They immediately wrote an abstract and addressed the unusual and disturbing findings in a presentation that was to be discussed at a meeting of the American Neurological Association in Chicago in 1954.

You might ask, what was known about the brain basis of memory in those days? In fact, the prominent theory of the time was championed by a famous Harvard psychologist named Karl Lashley, who had done a series of experiments in rats through which he tried to understand how memory was organized in the brain. He first taught the rats a maze and then systematically damaged different parts of the outer covering of their brains, or cortex, to see which area, when damaged, would lead to the most severe memory impairment for performance in the maze. What he found was that the location of the damage did not seem to make any difference. Instead he found that only when he damaged enough of the cortex did he see a memory deficit emerge. Based on these findings he concluded that memory was not localized to any particular part of the brain. Instead he believed memory was so complex that a large cortical network was involved and only when you damaged a significant part of that network would the memory system fail. This prominent view at the time made the striking memory deficits that Milner and

Penfield saw even more puzzling because the memory issues they observed seemed connected to the removal or damage of specific brain regions.

Several hundred miles away in Hartford, Connecticut, another neurosurgeon by the name of William Scoville read the abstract Penfield and Milner had submitted for the American Neurological Association conference and immediately contacted Penfield. Scoville had been treating a young man with such severe epilepsy that he, with the consent of the patient's family, had decided to do what he referred to as a "frankly experimental operation." Scoville removed the hippocampus and amygdala on both sides of the patient's brain, not just one. Scoville was correct about the reduction of epileptic seizures that took place, but immediately after the patient woke up it became clear that he, like Penfield and Milner's patients, had a profound memory deficit. He didn't know it at the time, but Scoville's patient (H.M.) was to become the most famous neurological patient ever studied.

Remember, this surgery took place at the height of the era when neurosurgeons were using brain operations—such as frontal lobotomies and procedures that damaged parts of the frontal and temporal lobes—to cure various psychiatric diseases like schizophrenia and bipolar disorder. This practice is referred to as psychosurgery. It's difficult to imagine what the mind-set was like at that time to believe it was okay to experiment with taking out parts of people's brains, even if it was supposed to be for their own good.

Scoville not only attended the 1954 meeting of the American Neurological Association but also presented a paper describing his patient H.M. Scoville then invited Milner to Connecticut to study patient H.M. She immediately jumped at the opportunity.

Milner has described herself as a "noticer," and her observations and testing of patient H.M. and nine other of Scoville's patients helped reveal something completely radical in our understanding of how memory works in the brain. H.M. was the easiest to test and evaluate because most of Scoville's other patients suffered from various psychiatric disorders, including schizophrenia and bipolar disorder. While she found H.M.'s intelligence to be quite high (and even improved a small amount after his surgery), he had a profound inability to remember anything that happened to him. He could not remember any of the hospital staff or doctors he came in contact with at the hospital (including Milner herself), could not find his way to the bathroom in the hospital or remember the location or address of the home his family moved to after his operation. Despite this profound inability to remember anything new, he knew his parents and the layout and location of his childhood home, and had apparently normal memories of his childhood. This meant that the operation carried out on H.M. impaired his ability to lay down new memories but spared his general intelligence (for example, he still continued to enjoy doing crossword puzzles—though he could do the same one over and over) and generally spared his memories of events that occurred before his operation.

This showed that Lashley's theory was wrong: There is a particular part of the brain that is specifically involved in allowing us to create new memories. What part of the brain was it? This is where Scoville and Milner were appropriately cautious. H.M.'s operation damaged both the hippocampus and the amygdala on both sides of the brain. This general region of the brain that houses the hippocampus and amygdala is often referred to as the medial temporal lobe. That is, it's the part of the brain's temporal lobe located toward the middle of the brain (*medial* in anatomical terminology). But in examining the nine other psychiatric patients who had varying amounts of medial temporal lobe damage, Scoville and Milner noticed that the more the hippocampus was damaged on both sides, the more severe the memory deficit. This led them to suggest that the large extent of hippocampal damage on both sides was likely underlying the severe memory deficit in H.M.; however, they could not rule out the possibility that concurrent damage of the amygdala plus the hippocampus was at the root of the memory loss.

THE FASCINATING STORY OF PATIENT H.M.

Patient H.M. is one of the most fascinating and most extensively studied neurological patients in the learning and memory literature. After Brenda Milner's work with him, her then graduate student and now professor emerita at MIT Suzanne Corkin studied H.M. for a total of forty-seven years, until his death in 2008. If you want to know more about patient H.M. and his story, I recommend Corkin's wonderful book, *Permanent Present Tense: The Unforgettable Life of the Amnesic Patient H.M.* Listen to me interview Suzanne about H.M. on the podcast *Transistor* by PRX.

But this was not all Milner noticed. Once she characterized the severity of H.M.'s everyday memory loss, she got to work figuring out if there was anything at all he could learn and remember normally. She and others later showed that H.M. had an inability to form any new memories for facts (termed semantic memory) or events (termed episodic memory), typically referred to together as declarative memory— the kind of memory that can be consciously brought to mind. Next Milner revealed that H.M. did have normal memory for some things. Namely, she showed that he still had the ability to learn new motor or perceptual skills at the same rate as people who had not undergone surgery. Milner had him do tests in which he had to learn how to trace a figure accurately while looking in a mirror. H.M. improved steadily day by day but, strikingly, had no memory of ever having done the task before. Similarly, he was able to learn perceptual tasks in which he was given a vague outline of a picture and, after a variable amount of time looking at the incomplete figure, gradually picked out the image. He learned to identify those objects at the same rate as nonpatients as well. This was another revelation in the memory field. This finding suggested that different brain areas outside of the hippocampal region were necessary for these forms of motor and perceptual memory.

So the partnership of Scoville and Milner revolutionized the way we understand memory. Their studies led to our understanding that the medial temporal lobe, which includes the hippocampus, is essential for our ability to form new memories for facts and events. The researchers also showed that memories are not stored in the hippocampus because H.M. retained normal memories of his childhood and demonstrated that different forms of memory, including perceptual memory and motor memory, depend on different brain areas outside the medial temporal lobe.

But one additional contribution of that original Scoville and Milner paper cannot be overlooked. The report served as a grave warning to the neurosurgical community that the bilateral removal of the hippocampus should never be done again. H.M. lost his ability to form any new memories and spent the rest of his life depending on his family to care for him. The operation took away his ability to retain anything new about what happened to him and what was going on in the world. It was a terrible price to pay for the reduction of his epileptic seizures, and Scoville and Milner made sure that the entire neurosurgical community understood.

DIFFERENT KINDS OF MEMORY

The kind of memory that H.M. lost with his brain damage is called declarative memory, which refers to those forms of memory that can be consciously recalled. In addition, there are two major categories of declarative memory that depend on the structures of the medial temporal lobe:

- **Episodic memory,** or memory for the events of our lives, which are those memories of our favorite Christmas celebrations or summer vacations; such "episodes" make up our unique personal histories.
- **Semantic memory,** which includes all the factual information we learn throughout our lives, such as the name of the states, the multiplication table, and phone numbers.

We now know there are many forms of memory that do not depend on the medial temporal lobe, for example:

- **Skills/habits:** These are the motor-based memories that allow us to learn to play tennis, hit a baseball, drive, or put our keys in our front door automatically. They depend on a set of brain structures called the striatum.
- **Priming:** This describes the phenomenon that exposure to one stimulus can affect the response to another stimulus. For example, if you give someone an incomplete sketch of an object that she can't identify but then show her a more complete sketch of the image, on the next round, she will be able to recognize the object even if less information is provided. Many different brain areas participate in priming.
- **Working memory:** This form of memory has been called the mental scratchpad and helps us keep relevant information in mind where it can be manipulated. For example, you are using your working memory during a talk with your financial adviser, who is describing the different mortgage rates for you as you try to decide which one is best for your situation. The ability to keep the figures in mind and manipulate them to come to a decision is an example of working memory. That H.M. could keep topics in mind enough to have normal conversations showed that his working memory was intact.

FINDING MY PLACE IN THE MEMORY MYSTERY

The groundbreaking report of Scoville and Milner in 1957 cracked the study of memory wide open and started an avalanche of new questions for neuroscientists to explore. Two questions at the top of the list were, first, figuring out which exact structures in the medial temporal lobe were critical for declarative memory: Was it just the hippocampus or the hippocampus and the amygdala? And, second, how do you visualize the specific change that occurs in the normal brain when a new declarative memory is formed? I didn't know it when I first began graduate school, but I was going to tackle the first question as my graduate thesis and the second question when I was an assistant professor at NYU.

By the time I entered U.C. San Diego in 1987, we knew a lot more about the important contribution of the hippocampus to memory, but the raging debate at the time focused on whether it was damage to the hippocampus alone that was underlying H.M.'s deficit, as Scoville and Milner hypothesized, or if it was the combined damage to the hippocampus and the amygdala, another possibility that could not be ruled out. A benchmark finding in animals in 1978 by Mort Mishkin appeared to provide evidence that it was the combined damage to both the hippocampus and the amygdala that led to the most severe memory deficits. Yet, in 1987 when I entered graduate school, Squire and Zola-Morgan at U.C. San Diego were finding evidence that the amygdala might not be involved after all. They had shown in animals that damaging the hippocampus on both sides caused a clear memory deficit, but they found no deficit after damaging just the amygdala on both sides of the brain. Then they did what turned out to be a key experiment. They added very precise damage to the amygdala in animals that had both their hippocampi removed. The researchers saw the addition of the selective amygdala damage did not in fact make the memory deficit worse, as predicted. The question was, If the additional memory impairment was not due to damage to the amygdala, then damage to what brain structure was it due to? A clue to this mystery came from a careful examination of the anatomy of the brain lesions. Neuroanatomist

David Amaral was looking at the extent of damage in the brains of these animals in thin sections of tissue and noticed something obvious only to a neuroanatomist: There was a lot more damage than to just the hippocampus and amygdala. Namely, a lot of the cortex surrounding the brain areas of these animals was also damaged, in varying degrees. It was likely that the same damage would be present in patient H.M., given the surgical approach used to make his brain lesion. Maybe the nondescript cortical areas surrounding the hippocampus and amygdala that nobody had ever considered very important, and had previously thought to be part of our visual system, were the key to the mystery.

This is where I entered the picture. Amaral ran a neuroanatomy lab at the Salk Institute in San Diego right across the street from U.C. San Diego, and he was a leading expert on the anatomical organization of the medial temporal lobe. It seemed clear to me that we needed a more careful understanding of the basic structure of this part of the brain, so when they asked me if I wanted to take on that challenge, I jumped at the opportunity. I literally felt like a neuroscientist version of David Livingstone, entering one of the deepest, darkest parts of the brain—somewhere few others had gone before.

I had thought for sure that all parts of the brain had been carefully examined and mapped in 1987 when I entered graduate school, but I soon found out that the areas I was focusing on had fallen through the cracks. I was one of the first to study them carefully. I used some of the same basic techniques that had been used by neuroanatomists since the early 1900s. I examined thin slices of the brain from key temporal lobe areas and stained them with a chemical to show the size and organization of the cell bodies of the neurons and glia that made up the tissue (this technique is called a Nissl stain). I looked at some slices to see if I could identify features that would allow me to differentiate one area from the next. In other studies, I tracked where these areas received inputs from and where they projected to.

I spent hundreds of hours over six years sitting alone in a darkened room staring at brain tissue under a powerful microscope trying to discern a clear pattern. Some days, after hours and hours of looking in the microscope at the cells that made up these brain regions, the images started dancing in front of my eyes like beautiful abstract pieces of art. It was hard, detailed work. Often, to fill the silence, I listened to classical music. Saturday mornings were my favorite microscope days. Sitting in the lab all alone in the dark with my slice of brain tissue I listened to a radio program called *Adventures in Good Music with Karl Haas*—a wonderful show from which I learned about everything from the mysteries of how Stradivarius made his famous violins to subtleties of the violin passages in Mendelssohn's symphonies. *Adventures in Good Music* was followed by the Metropolitan Opera's Saturday matinee, which broadcasts operas in their entirety. I should have gotten an additional PhD in classical music appreciation with all the hours I spent listening to these programs during graduate school. I had no human company during the time I spent in that dark room, but at least I had the music.

What did all this work tell me? It turns out that the cortical areas in the medial temporal lobe I studied, called the perirhinal and parahippocampal cortex, provide massive input into the hippocampus via a structure called the entorhinal cortex. In addition, my studies showed that these cortical areas are a major brain interface, or "gateway," receiving input from a wide range of brain areas involved in all kinds of sensory functions and other higher-level brain areas important for things like reward, attention, and cognition. Far from being simple visual areas, as researchers had previously thought, these regions are where high-level information converges in the brain. While I used relatively old-fashioned research approaches, my work revealed new information about why these brain areas might be so important for memory. Their *connections* were the key.

But just characterizing the connections of this region could not tell us exactly what its functions are. I went on to show that damage limited to these mystery cortical areas in animals causes devastating memory impairment that is similar in severity to the magnitude of impairment seen in H.M. This was another shocking finding. All the attention on memory research so far had been focused on the hippocampus and the amygdala. These new studies showed that neuroscience had been missing a key player in the game all along—the cortical areas that surround the hippocampus and amygdala. It was also clear that just because we implicated selective cortical areas important for memory, that did not mean that Lashley was vindicated. He had proposed that memory emerges from a complex interaction from widespread cortical areas across the brain and that no single area can underlie memory function. My findings showed that, in fact, you can identify specific and highly interconnected areas critical to the ability to form new long-term memories: specifically, the hippocampus and the cortical areas that immediately surround these structures. While Lashley was wrong about the localization of brain areas important for the formation of new memory, his ideas about the importance of large networks of brain areas did foreshadow findings that long-term memories can be stored in the same widespread cortical networks that process the incoming information in the first place.

My graduate studies helped identify two new brain areas and showed exactly how important they are for long-term memory function. In addition, the studies also pointed to another brain area, sitting between the perirhinal and parahippocampal cortices and the hippocampus, called the entorhinal cortex. Research shows that this area also plays a big part in the system of brain areas critical for declarative memory. Indeed, the recent Nobel Prize in Science or Medicine was awarded to two colleagues from Norway who characterized a major function of the entorhinal cortex in the processing of spatial information.

The U.C. San Diego research team and I hypothesized that patient H.M.'s severe memory impairment had to have been due to the damage both of the hippocampus and of *these surrounding cortical areas*. And sure enough, as soon I completed researching and writing my thesis, a brain scan was taken of H.M., which allowed researchers to visualize for the first time the true extent of his brain damage. This historic MRI scan (this is a technique that allows brain structure, including differentiating white matter, or axons, from gray matter, or cell bodies, to be visualized) confirmed that H.M. did sustain damage: not only to the hippocampus and amygdala but also to the surrounding cortical areas. This scan validated all the work that I had done for my dissertation, for which I earned my PhD and was awarded the prestigious Lindsley Prize, given by the Society for Neuroscience to the best doctoral dissertation in the field of behavioral neuroscience.

While I never met patient H.M., I thought so much about his brain and about what he could and could not remember that I felt like I *knew* him. I'll never forget the morning of December 4, 2008, when I opened up the *New York Times* to see his obituary on the front page. My first shock was to learn his full name for the first time in the twenty years I had been studying him. Henry Molaison. This was very likely the best-kept secret in all of neuroscience, revealed only at the time of his death. It was like learning something precious and very personal about a friend the day that he died. I happened to be teaching a big lecture course that day on the topic of memory. I shared the news with the class and even got a little emotional as I told them. They must have all thought I was a bit strange, but I couldn't help it. Henry Molaison, patient H.M., had given up so much in his life for our understanding of memory. Since the day of his surgery, he could never remember another Christmas or birthday celebration or vacation—he couldn't have a deep relationship with another person or make any plans for his future. He lost something precious the day of his surgery, but his misfortune enriched our knowledge of the brain and memory in a profound way. I will always honor his sacrifice.

MRI

MRI stands for "magnetic resonance imaging," and it is a powerful and common imaging tool that uses strong magnetic fields and radio waves to form images of the body, including the brain. This general imaging approach, also called structural imaging, is widely used to see the gross structure of the brain and the boundary between the so-called gray matter (cell bodies) and white matter (axonal pathways) of the brain.

MOVING ON: STUDYING MEMORY AT NIH AND STARTING MY OWN LAB

I had spent six years at U.C. San Diego mastering neuroanatomy and behavioral approaches to examine the connections of key brain areas in the medial temporal lobe as well as the effects of damage to those areas. While these are important areas of study, they still don't let you look firsthand at what's happening in the brain during the formation of new memories. That's what I wanted to do next. I wanted to learn new approaches by which I could examine the patterns of electrical activity in brain cells as animals performed different memory tasks. I wanted to look directly at the cells and the activity in the hippocampus that was occurring as animals learned something new. I secured a position as a postdoctoral fellow in the lab of Robert Desimone at the National Institutes of Health to do just that.

Desimone's lab was in the larger laboratory of neuropsychology run by Mort Mishkin, the same neuroscientist who had published key findings on the effects of hippocampus and amygdala lesions in animals and whom I had first heard about while in France. I spent the next four and a half years at NIH learning how to record the activity of individual and small groups of living brain cells as animals performed various memory tasks. This general approach is called behavioral neurophysiology, and it's powerful because we can examine how patterns of electrical activity in the brain relate to actual behavior. It is also powerful because it gives us a direct window on understanding exactly how particular brain cells respond to a given behavioral task. This contrasts with the studies of what happens with brain damage, like in the case of H.M. While transformative for our understanding of memory, lesion studies are, by their nature, indirect. We are studying the lack of function that used to be there before the damage. By contrast with behavioral neurophysiology, we can start to understand how the normal brain typically responds during a memory task.

It's important to note that there are no pain receptors in the brain, so the microelectrodes we used for the recordings didn't cause any discomfort, but they did allow us to record the brief electrical bursts of activity (called action potentials or spikes) that occur as an animal is learning or remembering something new. I basically trained animals to play video games focused on learning and memory and then recorded the activity of individual cells to figure out how the brain signals different aspects of the task and what happens to the pattern of brain activity when the brain remembers or forgets. I focused on one of the cortical brain areas in the medial temporal lobe, the entorhinal cortex, and characterized the patterns of neural activity in this area as animals performed a memory task. This was one of the only studies like this done in the entorhinal cortex. But I knew that there was much more left to understand relative to the physiological response properties of other key areas of the medial temporal lobe. That's what I wanted to focus on in my own lab.

Those four years at NIH were intense and very valuable because they taught me the ins and outs of this powerful approach
of behavioral neurophysiology, which I brought with me when I
started my own neuroscience research lab in 1998. This is where
things really got interesting in my career. I had at this point been
studying memory for ten years. I was thrilled beyond belief to now
be able to build my own research program focused on my scientific obsession—understanding what happens in the hippocampus
when a new memory is first formed. My desire to learn this was
inspired directly by the original description of patient H.M. He
could appreciate the things around him in the present moment, but
unlike us, he could not make that information stick in his brain
longer than he could focus his attention on it. We knew that the
ability to retain it depends on the hippocampus and surrounding
cortical areas, but we had no idea what these cells do when a new
memory is formed. *That* was the question that I wanted to investi-
gate in my lab.

So as head of my own lab, the first decision I needed to make was
what kind of information I was going to have the animals learn. It
had to be something relatively simple, so they could do it easily, and
be a task we knew was impaired by damage to the hippocampus and
surrounding structures. I settled on something that required animals
to associate particular visual cues (such as a picture of a dog or a
house or a building) with a particular rewarded target to the north,
south, east, or west on the computer monitor. We knew this form of
memory, called associative learning, was a subcategory of declarative
memory (in other words, it could be consciously learned and brought
to mind), and there was good evidence that damage to the hippocampus and/or its surrounding brain structures caused significant impairment in learning these picture–target associations.

I set about teaching animals to learn multiple new associations
each day, and when they could do the task very well, I introduced a

thin electrode into their brain to record activity as they were in the process of learning.

Finally! I was going to be able to peer into the brain and see what happens in the hippocampus as we learn something new.

One reason people had not done this kind of experiment before is because it's difficult to get animals to learn new associations. It turns out that the task that I chose was a good one; animals could learn multiple new associations in a given session. This was exactly what we needed to start looking at how new associations are signaled in the hippocampus.

Recording the activity of individual cells in the brain is a little like fishing. First, you set yourself up in a good part of the lake (or brain) where you think there will be some nice big fish (or brain cells), and then you wait. I was recording with a very thin microelectrode as it passed hundreds, probably thousands of cells through other parts of the brain before it reached the hippocampus. I sampled the electrical activity of the brain cells as the electrode passed by, and the brief burst of electrical activity registered as little "pops," which you can hear on an audio monitor. My goal was to figure out if the pattern of this firing from a given cell had anything to do with the animal learning a new association between a picture and a reward target. But there were no guarantees. There were plenty of days when I listened to the activity of many cells with the electrode, and not a one did anything much at all. It just sounded like a bunch of radio hash with no rhyme or reason to the pattern of firing. Other days, however, I got lucky and caught a nice big fish in the form of a cell that had interesting activity—for example, cells that seemed to fire only when a particular picture was shown or cells that fired a lot during the blank delay interval of the task between the presentation of the picture and when the animal made a response to one of the targets.

I kept fishing in the hippocampus with the hope of finding something interesting, and over the first few months of recording something did start to emerge. I noticed that a particular cell we were monitoring seemed to have little or no firing associated with the task early in the trial when the animal had not learned any of the associations. But then, the cell seemed to increase its firing rate later in the session when more associations were learned. I didn't fully appreciate the pattern until we analyzed the data later. Then it was as obvious as the nose on your face.

Just as I had noticed when listening to the cells during the experiment, these cells had little or no specific firing related to the task early in the learning session when the association had not been formed. But as the animal learned a new association, certain cells would dramatically increase their firing to double or sometimes triple their earlier rate. The increase in activity didn't happen when all associations were learned, just during particular associations. This suggested that there were particular groups of cells in the hippocampus that signaled the learning of new associations by changing their firing rate. I realized I had been listening to the birth of a new memory in the firing of these neurons! Nobody had ever characterized learning in the hippocampus in quite this way before. We were seeing exactly how hippocampal cells encoded newly learned associations, and because we know that damage to this brain region impairs the creation of such associations, the study suggested that this pattern of brain activity was the key to the new associative learning process.

This was not only exciting for my research partners and me but for the field of neuroscience in general. Ours was one of only a handful of demonstrations of brain plasticity occurring in real time and directly associated with a change in behavior, in this case, new associative learning! Diamond had shown that rat brains had more synapses in general if the animal was raised in an enriched environment relative to an impoverished environment, but her studies did not measure behavior while learning was occurring or while a memory was forming. It was kind of assumed that if your brain got bigger, this was generally a good thing for behavior or performance. The long-term implication of our work is that if we understand this functionality in the brain, we might be able to replicate it when a brain is handicapped by various neurological problems. In other words, these findings showed us how cells in a normal hippocampus work as new memories are being formed. Because this brain area was missing in H.M., he was not able to form new memories. Importantly, *these results are also key first steps to developing possible therapies for the associative or episodic memory deficits that occur in Alzheimer's disease, traumatic brain injury, and normal aging.* We must understand how the normal brain works to form new memories before we can fix what goes wrong with them in these neurological conditions.

TAKE-AWAYS: FORMING NEW MEMORIES

- Parts of the temporal lobe, including the hippocampus, entorhinal, perirhinal, and parahippocampal cortices (one on each side) are critical for a fundamental form of memory called declarative memory.
- Declarative memory is named for its ability to be consciously declared and includes memory for our life experiences (episodic memory) as well as memory for facts (semantic memory).
- For a new declarative memory to be laid down, these key temporal lobe regions must be working. These regions are also required as this new memory is repeatedly recalled and possibly associated with other information on its way to becoming a long-term memory.
- Once these temporal lobe areas do their job of forming a long-term memory, the areas are no longer required. The

memories are then thought to reside in complex networks of cells in the cortex.

- If you damage your hippocampus as an adult, there are no other brain areas that can take over. So there is no plasticity left if you lose this area of the brain.
- We now know that cells in the hippocampus can signal the formation of new associative memories by changing their firing rate in response to particular learned associations. Learn someone's name and there will be a group of cells in the hippocampus that are firing specifically to that newly learned name–face association.

THE MYSTERY OF MEMORY HITS HOME
Memories Mean More Than Neurons

It was a beautiful clear Wednesday morning in New York City. I had been a faculty member at NYU for a while by that time. That morning I was eager to dig into the Wednesday food section of the *New York Times*—my favorite section of the week. I was excited to see an article about the world-renowned chef Thomas Keller. With my parents, I had been to both of Keller's five-star restaurants, the French Laundry in Yountville, California, in the Napa Valley and the equally amazing Per Se, overlooking Central Park. I was looking forward to a fun article about specialty butter or rare wild mushrooms, so I was surprised when I found that the article was about Keller's later-in-life revived relationship with his father, who left the family when Keller was five years old.

Since that time, Keller had had only sporadic contact with his father. It was not until the chef was in his forties that son and father had established a real relationship. They enjoyed each other's company so much that the elder Keller moved to Yountville to live near his son. Both loved their new relationship, eating and enjoying life to its fullest—no doubt the wonderful food and the beautiful surroundings of the Napa Valley only added to the joyful intensity of their reunion. But a tragic car accident left Keller's father a paraplegic, requiring constant care and monitoring. Keller threw himself and all of his resources into helping his father heal and begin a new life from his now ubiquitous wheelchair. Under the careful watch of his son, the elder Keller survived with at least some of his old gusto intact for another year, before passing away.

It was a moving article. I could feel the pain that Keller experienced in losing the father whom he had only just gotten to know so late in life.

But the kicker was a quote by Keller who summed it all up when he said, "At the end of the day when we think about what we have, it's memories." I actually started to cry.

I cried not just because the story was moving. I cried because the story made me realize something important about myself. I had spent the last sixteen plus years studying the mechanics of memory, without truly thinking about what memories mean to me. Yes, I thought about patient H.M. and all that he lost without his medial temporal lobes. But I had spent no time thinking about how precious my own memories were to me. What memories were they? In a flash, the memories that came to mind were about studying; doing lab work; and earning degrees, prizes, and grants. I realized that my recent memories were all about science.

But I also had all sorts of memories about being a child, growing up in California with my parents and brother. If I concentrated, a slideshow of these moments began to unfold inside my head. Wasn't Thomas Keller right? Aren't our memories our most precious possession?

MEMORY LOSS AT HOME

While overt damage to the brain leading to amnesia is relatively rare, the brain regions damaged in those patients are also damaged in patients suffering from dementia and Alzheimer's disease. One January a few months after that Thomas Keller article had been published, I got a call from my mother. She told me that Dad wasn't feeling well and that he told her he couldn't remember how to get to the convenience store where he had been going to get his coffee for the last thirty plus years. Quite suddenly, my father's memory had evaporated.

I'm not a neurologist, but I knew my father's symptoms were not just the forgetfulness that comes with age, when the brain's memory centers begin to gradually slow down. I jumped into action and through my colleagues at Stanford, I got my dad an appointment with a top-notch neurologist. I flew out to go to the appointment with my parents. I was there when my father was given a diagnosis of general dementia.

I can't even put into words how helpless I felt. I was considered an expert on the brain areas important for memory, and yet I was completely and utterly powerless to do anything to help my dad. What was all my education good for if I couldn't help my own father? It was devastating.

I decided that even if I couldn't cure his memory problem, I would find a way to help him. In the process, I helped my mother and myself as well.

Before that fateful January, I had been on a mission (much like Thomas Keller) to improve and enrich my own relationship with both my parents. While we were never estranged, we were not close either. For many years, my parents and I spoke only once every few months. We had gotten into a habit of not speaking regularly. On my end, I was too busy trying to attain my dream of earning tenure as a professor of neuroscience. On their end, I think my parents just accepted the sparse communication as part of the package of having a daughter with the kind of high-powered academic career that had been expected of her.

But as I entered my forties, I decided that I wanted to close the distance between us. I first started by making a point to call every week, a change that they both embraced. So, I was already talking to Dad more regularly before and after his memory problem appeared. And after the diagnosis, Dad was still Dad: he had the same sweetness and sense of humor, loved to ask me if I had seen any good Broadway shows, and always loved to hear about the new restaurants in New York. He just couldn't remember what he had for lunch that day or who was at the family gathering last week.

But memory can also work in mysterious ways. Sometime after Dad's diagnosis, I decided I wanted to try to change another one of our family traditions. While my Japanese American family was unwaveringly polite and always friendly, one thing we were not was affectionate. I like to tell people to think of us like a Japanese American version of *Downton Abbey* without the accent, the servants, or the real estate.

While there was no question that my mom and dad loved my brother and me, the reality was that we never said it. That just wasn't done in the culture of our family. After we learned that my father had dementia, I realized that I wanted to start saying those three words, to both of my parents. I guess I wanted (or needed) them both to know that I *did* love them.

But then I had a problem. I knew I could not just start saying it without any explanation. It would be like suddenly starting to speak Russian to them instead of English for no reason.

So I decided I would have to ask permission.

Then I thought, Wait a minute! I'm a grown woman and I have to ask my parents' permission to say I love you? That's ridiculous, awkward, and uncomfortable. But then I realized that it wasn't the awkwardness of asking permission that was bothering me. It was the fear that they might say no. And I knew that would make me feel awful.

Well, the only way to find out was to ask. One Sunday night I gathered up all my courage before I made my regular call. Obviously, this was no ordinary Sunday night. This was going to be the night of "The Big Ask." How these Sunday-night calls worked was that I would first talk to my mom and share all my news of the week, and then she would hand the phone off to my dad and I would share all over again with him. I was starting to get scared that I would chicken out and not ask my question, so I decided my theme for the phone call would be "Keep it light." I would treat my request like any other request, like "Hey, Mom, what if we start talking on Monday nights instead of Sunday nights?" This was my strategy. It seemed better than "Hey, Mom, how about we try to change thousands of years of stoic and deeply ingrained Japanese culture in one fell swoop and start to say I love you to each other?"

The first part of the call was like any other Sunday phone call. I asked her how her week was, and I told her about mine. I was especially upbeat and cheerful that night and somewhere in the middle of the conversation I launched in.

"Hey, Mom, I realized we never say I love you on our phone calls. What do you think if we start saying that?"

There was a pause in the conversation.

A really long pause.

I think I was holding my breath. But when she finally answered she said, "I think that's a great idea!"

I gulped air and breathed a huge, silent sigh of relief.

Sticking to my theme of keeping it light, I replied, "That's *great*!"

We finished up our conversation about what we did that week, and I could feel a tension growing in our voices. We were like a couple of wild cougars warily circling each other. Why the tension? Because I think we both knew that it's one thing to *agree* to say I love you and a very different thing to actually *say* I love you for the very first time.

But it was my idea, so I took the bull by the horns and said, "Okayyeeeee" (in other words, get ready for it, Mom!).

"*I love you,*" I said in a big, overexaggerated Disney-like voice to hide my discomfort.

She replied "*I love you too,*" in an equally exaggerated voice.

I won't lie, it was very difficult and very awkward, but we did it! Thank goodness that was over!

I knew that once my mom agreed my dad would agree too. During my conversation with him that night, I asked permission, he said yes, and we said our awkward I love you's to each other, and the historic night of The Big Ask was over.

I should have been so proud and happy when I hung up from that call. And I was, but I also burst out crying when I got off the phone. The fact was, nothing about what had happened was light. I had said I love you to my parents for the first time that night as an adult, and they had said it right back to me. With that, we had shifted the culture of my family—forever. It was moving, and my tears were mostly tears of joy.

The following week, I was happy to see that saying I love you had already become much less awkward with my mom.

Then it was Dad's turn. I realized that there was a chance that he might not remember our conversation from the week before, so I was ready to remind him about our agreement.

But that night, Dad surprised me.

You see, that night, and every Sunday night conversation since, my dad has said I love you first. He remembered.

You have to understand that sometimes he can't quite remember whether I'm visiting for Thanksgiving or Christmas, but he remembers to say I love you at the end of every phone call without fail.

As a neuroscientist, I immediately recognized why this happened. This is a beautiful example of the power of emotion to strengthen memory. The love and maybe even the pride my dad felt the week before when his daughter asked if she could tell him she loved him—that emotion beat dementia and allowed him to form a new long-term memory that has lasted to this day. When events or information arouses us emotionally, our amygdala gets activated; that brain area, we now know, is critical for processing emotion and helps boost the memory processed by the hippocampus. This shows just how interdependent emotion and cognition, or feeling and learning, truly are.

That night, my dad formed a new long-term memory despite his dementia. And you can be sure, the memory of that phone call will be locked into *my* brain for the rest of my life.

WHAT MAKES SOMETHING MEMORABLE?

The fact that my father always remembers to say I love you at the end of our phone calls is an example of how emotional resonance can make memories stronger. But emotional resonance that kicks the amygdala into gear is not the only thing that can boost memory. For example, my request to my dad was also very novel relative to our other conversations in the more than forty years he has known me, and novelty is another key factor that can enhance memory. You see, our brains are naturally tuned in to novelty. It's actually a safety issue because we want to be vigilant of new things in our environment that might be dangerous. Our brains tend to respond strongest (in terms of action potentials) to new stimuli so that a bigger response will be seen when we are looking at a completely novel face, for example, instead of the face of our office mate whom we see every day. It turns out that novel information is also easier to remember.

But there are a few other key factors that improve our memory, which I notice in my dad every week during the football and base-ball seasons. You see, he can often tell me what football or baseball game he watched—especially if it was an exciting one. Just a few days ago, he told me he really enjoyed watching his San Francisco Giants win the 2014 World Series against Kansas City and that it was especially exciting because the Giants won in game seven. To tell the truth, I don't follow baseball and I had to google it to be sure he got his facts right. Now, *that's* pretty darn good memory for someone with dementia! The trick there is that my father *loves* baseball, especially the Giants, and has essentially a lifetime of memories and associations with Giants baseball that make it easier for him to remember the details of the World Series. All those associations that he has with the Giants provided a framework for remembering this new but associated piece of information: The Giants won the World Series (again) in 2014! We know that one of the major functions of the hippocampus is to help link or associate initially unrelated items in memory. The larger associative network is stored in the cortex, but when the hippocampus can link a new item (like the Giants winning the Series) to a much larger network of other Giants baseball–related information, it becomes easier both to learn and to remember that information. This is part of what makes my dad still my dad, even if his ability to form new memories is weaker now. He has a foundation of strong memory networks, which he has built up throughout his life, of the things he loves, thinks, and cares about: his family, food, Broadway, baseball, and football, to give a few examples. I am so thankful that this aspect of memory allows Dad to retain all the things he enjoys most.

DEFINING DEMENTIA AND ALZHEIMER'S DISEASE

How are dementia and Alzheimer's related? *Dementia* is a general term that describes a set of symptoms that are severe enough to affect a person's everyday life. These symptoms most commonly include a decline in memory function, planning ability, decision making, and other thinking skills. The term alone does not describe a specific disease. *Alzheimer's disease* is the most common form of dementia. It is estimated that 60 to 80 percent of people with signs of dementia have Alzheimer's disease. The most common symptom of Alzheimer's disease is difficulty remembering names and recent events. It is associated with deposits of protein fragments called beta-amyloid (referred to as plaque) and twisted strands of another protein called au (referred to as tangles). These plaques and tangles are found all over the brain in late stages of the disease. To find out more information see the Alzheimer's Association's website (www.alz.org).

MARRIED TO SCIENCE

Yes, I was trying to strengthen my relationship with my family, but my career was still the major focus of my life. My new appreciation of how precious our memories are was making me realize how few precious personal memories I had. Don't get me wrong. I had many great colleagues and work friends. Over the years I had established strong and productive collaborations for my work. I was considered an energetic and productive colleague to many but dear friend to few. Not to mention the fact that I was perpetually single. Was being married to science enough for me?

My focus on work and more specifically on succeeding in science was not new. It started when I was an undergraduate with my determination to become a neuroscientist and teacher like Professor Diamond. The irony here is that Diamond was more than a science role model. She was a wonderful role model for a balanced life in science that included not only an active research lab and spectacular teaching reputation but also a husband (another scientist), children, and an active social life that included her weekly undergraduate tennis matches. But for some reason, I didn't feel the need to model myself after those other features of her identity. All I focused on was her passion for research and professional success.

My focus on work was amped up even more once I started my postdoctoral position at NIH. Every day, typically seven days a week, I had a forty-minute commute from my apartment in the Adams Morgan neighborhood in D.C. to my small office in the basement of Building 49 on the campus of the National Institutes of Health in Bethesda, Maryland, to work, work, work. Yes, I did socialize with the great group of post-doc colleagues in the lab at the time, and even dated two of my colleagues (at different times), which was when I learned firsthand all the reasons people warn you *not* to date your coworkers. But these were momentary blips in the scheme of my life. I hadn't had a serious boyfriend since François.

I had developed a theory about myself that started to define how I lived my life. My theory was that my self-worth was measured only by how many papers I published and grants and prizes I won. It made a lot of sense at the time; certainly this was the area of my life where I got the most attention and recognition. It was also an easy formula to follow. It was easier just to work all the time. There were no messy emotional attachments to deal with—just doing the work to the best of my ability. Yes, I could do that, *and* I was really good at it.

But there were a couple of corollaries too. One was the idea that I was not good in purely nonscience social situations. I felt confident about how I handled social situations revolving around science. If I could talk about my passion for my work, I was in my element. The problem was I didn't know how to talk about anything else, which made my social conversation both awkward and boring. I had also decided during this time that men were just not interested in me. I had lots of great evidence for this theory. Just consider my graduate school experience. Six whole years and just one real date during all of that time. Actually someone else asked me out, but the first date went so badly, I said I was too busy and couldn't go out again. The men I dated from my own lab left me feeling alienated from my own work environment and even less enthusiastic about my dating ability. Yup, my theory was clearly correct: Men simply were not interested in me, and it was not worth the bother.

The first year I got to NYU, amazingly, I was asked to be photographed for Annie Leibovitz's photo-essay book about women called, appropriately, *Women*. And it was because of my teaching. I had been asked by my department to organize a day-long set of lectures for talented thirteen-year-olds who got great scores on their PSAT tests. Being very comfortable with human neuroanatomy after all my coursework with Marian Diamond, I decided to teach a section on human brain anatomy. The NYU newspaper had printed a photo of me holding a preserved human brain (just like the one Diamond first showed me) with a group of mesmerized teens looking on. Susan Sontag saw that picture (she was an adjunct teacher at the college), thought that I was an excellent example of an intellectual woman, and suggested that Leibovitz ask me to be in the book!

I didn't have to be asked twice, and before I knew it, Leibovitz was standing in my lab. I ended up on a full two-page spread between Frances McDormand on one side and Gwyneth Paltrow and Blythe Danner on the other side. Pretty glamorous, right?

Someone once commented after seeing that picture that I must have men lining up outside my lab door.

My response: "Ha!" This was not a very polite response to a very nice compliment.

The fact was that, although I was a bona fide Annie Leibovitz model, there were *never* any men lining up at my lab door, let alone my apartment door. See? Men were just not interested in me.

Despite my obsessive work life and clear lack of a social life, I allowed myself one pleasure: good food. Partially primed by my time with François in Bordeaux and genetically primed by my parents, I loved food, and the restaurant scene in New York was phenomenally interesting. I read all the restaurant reviews (hence my paying attention to the article about Thomas Keller) and listened for any buzz I could pick up on the best and most interesting restaurants around.

When I first arrived as a new assistant professor at NYU I happily agreed to organize the departmental speaker series for the year with a colleague. We took suggestions for speakers from the faculty, made the invitations, and were responsible for hosting the guests during their visit. But the real reason I loved this job was because I got to choose the restaurants we took the speakers to after their talks. I took full advantage of this opportunity by researching and choosing what I thought would be the absolute perfect restaurant for a particular speaker—whether I knew the speaker or not. Those evenings out were becoming virtually my only social outlet, so, as was my way, I threw myself into restaurant research with all my might.

I ate by myself at the bars of the most interesting restaurants that I could find in New York. I liked to try new restaurants, but I became a regular at several neighborhood places too. I knew I had become a *real* regular when the bartenders started comping my meal because I ate there so often. All of this food-centered "research" could lead to only one outcome: chunkiness. My own.

Soon after I received tenure at NYU, I learned that I had been selected to receive the Troland Research Award from the National Academy of Sciences. This yearly prize honors the best under-forty researchers in the area of experimental psychology in the country. It was an amazing and thrilling honor to receive this award. It was particularly special because my parents flew out from California to attend the ceremony in April 2004 at the National Academy of Sciences in Washington, D.C. My former dissertation adviser Larry Squire from U.C. San Diego, who is himself a member of the National Academy of Sciences, was also there, and he, my parents, and I all went out for a wonderful dinner to celebrate the occasion.

The author holding the Troland Research Award with her parents at the National Academy of Sciences in Washington, D.C.

Yes, I am smiling in the picture, and I was indeed happy that night. But beneath the smile was a woman who, at thirty-nine years old, was becoming aware of something: *She hadn't really given a moment's thought to anything but her career and neuroscience experiments for years and years.* And when I tentatively popped my head outside my lab door to take a look at New York City, I found myself all alone. It was as if I were leading a double life. My science life was like one big party that you never wanted to leave, with lots of engaging colleagues to talk to and always something new and interesting in the works. By contrast, my social life was like one of those deserted ghost towns in a Clint Eastwood western with bunches of tumbleweed swirling around the dusty road. In my quest to push the limits of science in my field and gain the coveted position of tenure at NYU, I'd lost so much of myself.

You can see it in that picture from the Troland Research Award ceremony with my parents. I was getting wide. But my waistline was not the only thing changing. Something even bigger was starting to change for me. I had finally reached my goal. I had won tenure at NYU and had a big active research lab to show for it. Yes, I was pleased, but then again, I was a little lost. What was there to work for now that tenure had been achieved? I could make it to full professor status above my current rank of associate professor. But what else? I thought I would have *everything* once I got tenure. The truth is that I had a title and a great research program that I loved, but not much else.

Maybe I needed to pay more attention to Thomas Keller and start making some of the kind of memories that would really matter to me.

It was scary to consider these possibilities. Considering these possibilities meant that I had to admit how bad things were and I was not quite ready to do that. So many things were wrong at that point. How was I going to feel better about myself? How was I going to reconnect with the little girl who wanted to be a Broadway star? The romantic who fell in love with a French musician? Where had that woman gone?

I was going to find out, and I was going to use my brain to figure it out.

TAKE-AWAYS: WHAT MAKES THINGS MEMORABLE?

While we are still waiting for that magic pill that will allow us to magically remember exactly what we want and need to remember, here are some practical tips for how to make things stick in your memory.

- The more you bring a memory back to mind, the stronger it becomes. Boring but true. At the neural level, with each **repetition** you are strengthening the synaptic connections underlying the memory, allowing it to resist interference from other memories or general degradation. Repetition engages the neural networks related to our attention system; in other words, we tend to remember what we pay attention to.

- If you want to remember something new, try to link or associate it to something you already know well, and this will help. The more **associations** a memory has, the stronger it is because it allows the memory to be retrieved in the widest variety of ways. If one clue doesn't work, it will always have another to help retrieve it.

- We know that memories with *emotional resonance* last longer and are stronger than other memories. This is because the amygdala, a structure critical for the processing of emotion, has the ability to form very long-lasting memories with help from the hippocampus. From an evolutionary point of view, the amygdala (one of the oldest parts of our brain) signaled us in an automatic way whether something in the environment was good or dangerous. As our brains evolved into more complex structures, the amygdala started sending reinforcement to the hippocampus whenever it picked up salient emotional experiences. It signals to the hippocampus: Remember this moment, it made me laugh, cry, scream with fear! It's for this reason that our strong emotional memories seem imprinted on our brain and are so long lasting.

- The brain is wired to focus attention on *novelty* so really novel events—the only time it ever snowed when you were in California or the one time you saw a meteor shower—tend to be memorable.

BRAIN HACKS: FROM A MEMORY CHAMPION

I recently spoke at a TEDx event in the Bay Area and one of the other speakers was the 2008 U.S. National Memory Champion, Chester Santos. He dazzled everyone by reciting the names of probably eighty to ninety people in the audience that he had met briefly just that day. Then he did something even more amazing. He recited the following list of thirteen words quite quickly:

Monkey
Iron
Rope
Kite
House
Paper
Shoe
Worm
Pencil
Envelope
River
Rock
Tree
Cheese
Quarter

He told us that he could get us to remember that list of words in just about three minutes. We were all waiting with bated breath. Then he went on to use some of the key factors that make things memorable, including novelty, emotional resonance, and associations. He told us that one way to remember a long list of unrelated items is to make up a story with those items; and the more fantastical or funny the story, the more memorable it would be. Then he recited such a story for us. He started by asking us to picture a monkey pumping iron (a novel and funny image). Then a big rope descends out of the sky. Imagine yourself feeling the texture of the big rope. You look up and see that the rope is connected to a kite. But no sooner do you notice the kite than a huge wind comes up and blows the kite right into the side of a house. That house is covered with pieces of paper. Imagine a house covered with hundreds of yellow sticky notes tiled all over it. Then a gigantic shoe appears and starts walking around the house covered with paper, making shoe marks all over it. But this shoe is really smelly, so picture a little worm boring its way out of the inside sole of the shoe. Suddenly, the worm turns into a pencil and starts writing on an envelope that appears on the roof of the house. Another big wind comes up and the pencil and envelope are both blown into a raging river. Then imagine the river so raging that waves begin to crash onto a big rock. The rock turns into a beautiful tree, but this tree is unusual: it is growing cheese on it. And then the most striking thing happens—suddenly quarters start shooting out of the cheese on the cheese tree.

Okay, we all agreed it was a fantastical story. But how memorable is it? Santos then started to recite the story again with the entire audience (including me sitting in the front row) yelling out all the key words as he told the other parts of the story, and it was clear that his fantastical story with its improbable events really worked to help us remember! After that he asked us to recite back the list and the whole group of three hundred people simultaneously recited the list perfectly from beginning to end. Amazing! He gave us that memorable story. Clearly it will take practice to come up with your own fantastical story that will help you remember a long list of things. But the cool thing is that Santos was using the same tools that we know improve memory—association, emotional resonance (humor), and novelty are at play big time in his story—to speed and enhance the learning process. I'm a convert! I'm going to have to start practicing with my own fantastical stories next time I need to remember a list of errands to run or a list of points to make in a presentation!

4

CHUNKY NO MORE
Reconnecting My Brain with My Body and Spirit

There is often a defining moment in people's lives that makes them decide to change their habits and routines and get fit. A health scare, a class reunion, a particularly unflattering picture—any of these things might do the trick. I was sick and tired of being overweight, but I had only ever made half-hearted attempts to change my sedentary, foodie ways. It wasn't until I was on a white-water rafting trip in South America that I had the realization that gave me the motivation I needed to start the process of getting in shape.

A RIVER-RAFTING WAKE-UP CALL

It was in July 2002, and we were at the end of another gorgeous day on the mighty Cotahuasi River in central Peru. Mark, our fearless guide, was steering our raft. I was part of a group of fun-loving fellow adventure travelers, including a bunch of triathletes from northern California, a father–daughter pair, a river-loving husband-and-wife team, and Cea Higgins, a super-cool surfer and mother of two who became my rafting partner for the trip. I had come on this trip on my own searching for a little adventure and to get away from the relentless grind of science. We were all whizzing down the class-five river in the deepest canyon in the world surrounded by steep cliffs full of craggy gray rock formations. This trip was the latest in a series of adventure travel vacations that I had taken over the past few years, including a kayaking trip in Crete and another river-rafting trip down the Zambezi River in Zimbabwe the year before. I might have been living the life of a cloistered lab rat in New York City, but I made a point once a year to indulge the world traveler in me and let my hair down as far away from the hustle and bustle of New York City as I could. For me, white-water river rafting or kayaking in exotic places did the trick.

Even before we got on the river, this Peruvian adventure started with a six-hour bus trip from the airport at Arequipa, Peru, to the tiny town where we stayed overnight in a very rustic hotel before what was described to us as a "brutal" ten-hour hike on the actual Inca trail to where our boats were moored on the Cotahuasi River. I will never forget the ice-cold shower (adventure travel vacations don't always come with hot water it seems) I took the morning before we all set out, mules in tow, on the hike. It was a bright and glorious day, and with our constant chitchatting as we got to know each other, even the long hike was over before we knew it. We were all tired by the end but were happy to have found our way to our floating caravan of rafts securely tied to the bank of the river. I remember thinking that the way those rafts were bouncing up and down on the water, it looked as if they were as eager as we were to finally explore the river.

Each evening after a long day of rafting, our guides chose a campsite somewhere on the banks of the river. Each night in camp, our first job was to get all the camping equipment and all our personal bags up from the supply rafts to the campsite. To do this, we formed a human "fire line" and handed each bag or piece of equipment from one person to the next until it reached camp. Our fire line sometimes went up a steep incline from the river's edge.

It was that first night, standing somewhere in the middle of the line, that I received my personal, undisputed, loud and clear fitness wake-up call. Why? Because that night in the fire line was when I appreciated how truly pitiful my upper body strength was. At that time, I had several years of regular yoga under my belt that had shifted my totally inflexible body into a somewhat less inflexible body, but I had done virtually no strength or aerobic training and it showed. I now found myself the weakest link in our human chain. Not only was the sixteen-year-old girl there with her father much stronger than I was but there were sixty-five-plus-year-olds who blew me out of the water in terms of strength. Of course, my fellow river rafters never let the large packs being passed up to camp crush me like a bug. Instead, two fellow rafters got on either side of me and essentially passed the heaviest packs to each other while making it look like I was helping, thankfully allowing me to save face.

I was mortified. I literally could not pull my own weight.

That fact drove home a searing kind of shame: I was young, healthy, and able. Why could I not keep up with my fellow adventure travelers?

It was that night in the fire line that I made myself a promise: I was going to get in shape—get strong and healthy, nimble and quick—as soon as I got back to New York.

FIT, FAT, AND FEARFUL

True to my promise to myself on the river, two days after I returned from my rafting adventure, I marched myself down to a brand-new Equinox Fitness club that had opened up not too far from my lab. This gym was beautiful and had everything—a big facility with yoga and Pilates studios, workout rooms, personal trainers, fancy locker rooms, a sauna, and a pool. And it was only a fifteen-minute walk from work. It was perfect! I signed up immediately. The membership adviser sold me on trying out a personal trainer because a free training session came with my new membership package. I was determined to do this right so I marched straight up to the board on the wall with all the trainer profiles and carefully tried to choose the one who looked like he or she could get me into shape the fastest. Five days later, I had my first session with my new personal trainer, Carrie Newport.

Carrie was a relatively new trainer at the gym looking to build up her clientele. Turns out, she was the perfect trainer for me. Always bubbling over with information from the latest personal training seminar, she was enthusiastic, knowledgeable, creative in her workout design, and well organized. It was so much fun to train with her two to three times a week. The best part was that I quickly started to see results in both the increased weights I could handle and the higher number of reps I was able to complete in our sessions as well as in the shape of my body. Muscle mass grows if you work out regularly and if you push yourself hard. To supplement my training sessions with Carrie, I started taking advantage of the great fitness classes at the gym too. They had especially good dance teachers (there are so many fantastic dancers in New York and we get them as teachers in the gyms), and I also enjoyed the cardio, strength-training, and step aerobics classes. I tried them all!

When I look back on this time, I realize how many old habits I broke and how many new habits I established in one fell swoop as soon as I got back from that trip to Peru. All the books on breaking habits say it's so difficult to do because habitual behavior is ingrained and unconscious and therefore very difficult to change. But my sudden change from non–gym goer to regular gym goer didn't seem hard to me at all. Why? The first key factor was that I really had a profound revelation that night on the banks of the Cotahuasi

River. That realization opened my eyes for the first time about my fitness level, and I was determined not to again be the weakest one on any future trip, which completely shifted my motivation to work out. The second key factor that was critical is that on top of the expensive gym membership, I hired a regular trainer to work with me one on one two or three times a week. I like to get my money's worth, and that kept me extra motivated to get the most out of each personal training session. It was really Carrie who got me over the hump and helped jump-start my new habit of going to the gym on a regular basis. Her style of combining copious amounts of positive feedback and encouragement with fun and varied workouts along with a bubbly personality made the workouts so enjoyable—I loved them. The last key factor during the first year or year and a half after Peru was that I quickly started to see the fruits of my labor in terms of clear increases in my strength and changes in my body. That motivation alone was powerful enough to keep me going to my regular sessions with Carrie with no problem at all.

After the initial eighteen months of weight training and cardio with Carrie, I had reached my first fitness goal. I was much stronger and was ready to haul huge pieces of equipment on any fire line I might be asked to join. I had a much higher aerobic capacity and was ready to take any cardio fitness test Carrie could throw at me. I had also become a regular and enthusiastic gym goer; you could set your watch by my regular visits to the club, and I was on my way to becoming a bona fide gym rat.

But, despite all these positive changes, that still did not mean I was totally fit. While I was significantly stronger in 2004 than I had been in 2002, the truth was that I was still overweight and had even started to gain more weight in 2004. There were two main reasons for that. The first obvious reason was my eating habits. I was still indulging in great restaurants and the best take-out I could find in the city. My regular visits to the vending machine on the first floor of the building (better to hide my terrible habit from the people in my lab and in my department who had their offices on the eighth and eleventh floors), where I indulged in a Twix bar, my very favorite candy bar, before most every workout. That combination of chewy caramel together with a crunchy cookie on the bottom all covered in chocolate was irresistible and made me feel like I was gathering my strength and energy for my upcoming workout with Carrie. So while I was much stronger and firmer, I was still carrying too much weight on my five-foot, four-inch frame.

I started to realize that I needed to pay attention to what and how much I ate even beyond my Twix bar habit. It was the height of the Atkins and South Beach diet crazes, and this got me thinking about how many carbs I consumed every day—in fact at all meals, all day long. For example, I loved to make my own waffles for breakfast in the morning and eat those freshly made waffles with butter and syrup. Not just Sunday mornings, but every morning. I could make the fastest and most delicious fresh homemade waffles in the city, and they immediately went from my plate to my hips. When I got tired of waffles, I would go out and find a wonderful hunk of peasant, walnut, or date bread and toast and butter that for breakfast. You can't imagine the number of amazing bakeries in New York, with bread so much better than any bread I ever had in California or Washington, D.C. Yum! Lunch was often a sandwich on good bread, and dinner came from a fantastic array of restaurants (lots of pasta included) or take-out in my Greenwich Village neighborhood. One of my very favorite meals was bulgogi, a Korean barbecued beef dish with noodles from a fantastic pan-Asian restaurant in Soho. The dish could have easily fed two or three people. I ate the whole thing myself for dinner with a side serving of rice to sop up all that delicious sauce—heavenly! With this kind of eating lifestyle, it was no wonder I ended up at least twenty pounds over-weight despite my regular gym habit.

Fit, fat, and fearful. Those were the three words that best de-scribed Wendy Suzuki in 2004. The "fearful" part was mainly due to another major factor affecting my life at that point. I was smack dab in the middle of that inevitable trial by fire for academics: win-ning tenure. Here is a Cliffs Notes version for the uninitiated. First, you are lucky enough to be hired by a big fancy research university that gives you a shiny new lab and a pot of money that is just big enough to get your groundbreaking neuroscience research going but not enough to sustain it for more than a couple of years at the most. This happened in 1998 for me. As soon as you arrive in your new home, you immediately start setting up your brand-new lab and at the same time you start madly writing as many grants as you possi-bly can to increase your chance of being funded so your lab doesn't fold after just a couple of years when your startup funds run out. Oh, and at the same time you also have to start teaching classes, mentoring graduate students, and hiring technical staff, most of which you have not done much of before because you were too busy doing the science experiments that got you the job.

From the time you are hired, you typically have six years to show your stuff in terms of hard-core research publications in peer-reviewed journals, teaching, and mentorship, though at major research institutions it's the research productivity measured in terms of high-profile publications that your senior colleagues (the ones who vote on your tenure) are really interested in. That means you have only six years to fund your lab (using hard-to-get big-government grants for which you are competing with Nobel Prize winners), get your experiments working, and find something really earth-shatteringly interesting to publish to great acclaim. Some-times it takes years to simply set up a lab, depending on the kind of experiments you plan to do. Of course, mine were the kind that needed a lot of setup.

If that pressure doesn't sound intense enough, the worst part of the process is that no matter how many papers you publish or great classes you teach, you are never 100 percent sure if you have done enough to earn that lofty status of tenure. Inevitably you start hearing about the exceptions. As in, this superstar neuroscientist from a prestigious university who mysteriously did not get tenure and nobody quite knows why. You'll hear, "He was clearly an exception; you will do fine." And you immediately think, "What if I'm one of those exceptions too?"

I fully admit that I was an anxious, agonizing, agitated, and stressed-out assistant professor trying my best to make tenure in 2004. I was working for what I hoped would be a spectacular result from a challenging experiment that took several years to develop. In the end, it worked out fine, but I had many a sleepless night worrying about the speed at which my research was progressing and how interesting my findings would be to my peers.

I waited until after that positive tenure decision came down to really dig my heels in and tackle those excess twenty pounds with better control of my food portions and a drastic reduction of carbs in my diet. While my motivation to get fit was my Peruvian river-rafting adventure, my motivation to lose weight was that picture of me with my parents at the Troland Research Award ceremony in Washington. In 2004, I was plenty fit, but my outside body didn't reflect my inner strength, and that just didn't feel right. Inspired by how my body had immediately responded to my strength training and cardio routine as designed by Carrie, I decided I was going to tackle my food issues solo. How hard could it be?

I started by planning all my meals and took careful note of my portion sizes. I researched fun and easy recipes to cook to adhere to my new diet and significantly decreased my take-out and restaurant going, except for special occasions. The cooking part was fun, but the biggest challenge was to get comfortable with a feeling of hunger—all the time. I hated it. I especially hated it late in the afternoon just past the midway point between lunch and dinner, when I would start thinking about heading down to the vending machine to get that crispy, gooey Twix bar. I got grumpy. I got cranky. I could not concentrate well and started to tell myself that eating a sugary snack was okay because it would make me more productive at work, and I needed to be maximally productive at work. But I had set my food rules, and I muscled through.

This was particularly hard because my weight loss seemed to go so much slower than my muscle buildup. It took many weeks to start to see an effect on the scale. But there were results—slow but steady. When I saw those first two or three pounds disappear from the scale, I felt newly motivated. That made me realize that the feeling of being hungry that I hated so much was actually *good* and it translated to *results*, however slowly. That hungry feeling was also helping me change my body's set point for food intake, for the better. What do I mean by that? It probably took me years to get myself to the level of food intake that my body needed to feel satisfied, without overeating. I realized my diet was too high in calories and too full of carbs, without enough veggies or fruits. I knew it was going to take time and some determination to slowly ramp down as I started to gradually change the composition of what I ate every day. I got more creative with my cooking, looking up healthful low-carb recipes online, and the truth was, with my new focus on more nutritious cooking, I didn't even miss my old unhealthy take-out regime. My trainer had shown me that I could strengthen my body with regular smart workouts. Now I was learning how to get my eating, and therefore my weight, back in control with the same kind of slow, steady, and smart food choice changes.

EXPLORING THE FOOD SET
POINT IN . . . MY CAT

Experiencing hunger but then seeing the number on my scale go down as my own eating set point changed had another unexpected positive benefit—with my overweight cat, Pepper. Pepper has no brakes on his brain's hunger center but his brother, Dill, before he passed away, was always very thin and a picky eater. To make sure Dill ate enough, I would leave food out all the time. I was so worried about Dill getting what he needed that I didn't notice Pepper's expanding middle until friends started commenting on the size of my cat! Well, I realized, a little too late, that Pepper (like his human, just a few years before) was in great need of a diet, and I knew what I had to do. I put him on a very regimented (and vet-approved) low-calorie diet on which he got fed only twice a day with special low-calorie cat food. He was uncomfortable at first. Very uncomfortable. He was *so* hungry, he would start running around like a mad man before mealtimes, but I knew from the vet that he was getting proper nutrition, and his body just had to get used to this amount of food, and then his brain would settle down. Like me, he had to change his set point. I didn't waiver and just kept to his feeding schedule for weeks and weeks. I could see not only his waistline slowly whittle but his set point shift as well. Early on he would inhale his food and lick the empty bowl. Now, for example, he leaves about a third of the food for a snack later. He always finishes, but he can now spread out his eating in a leisurely way. I sometimes wished someone would just feed me the prescribed amount at the prescribed time too. The greatest thing to see was Pepper's energy skyrocket (just like mine did) when that massive waistline started to get smaller. He still runs around the apartment—but for fun now, not because he is crazy hungry.

I remember the day I went to a jazz dance class and a guy whom I danced with pretty regularly for at least a year but had not seen for a few months did a double take when I walked in. He said he hardly recognized me because of my weight loss. That right there was all I needed to hear to know that all those months of that hungry feeling were totally worth it. Talk about immediate gratification. I was ecstatic! Over about a year of combining my new diet with my regular gym workouts (by this time I had graduated from Carrie's care and did my own workouts, mainly focusing on different classes at the gym) I ended up losing a total of twenty-three pounds. Woo-hoo!

That should have been enough, right? Plenty even. But there was something even more in store for me.

DISCOVERING A WORKOUT WITH A MESSAGE

During one of my regular evening sessions at the gym when I had already attained most of my weight-loss goal, the list of possible classes caught my eye. I had a choice that evening between a cardio boot camp class and another class that I had never heard of called intenSati—with no explanation for what *intenSati* meant. I was not feeling all that energetic, and the cardio boot camp class just sounded too hard. So that's how I ended up walking into my first intenSati class. Little did I know that this class was not only harder than cardio boot camp but would be the catalyst for upping the level of my workouts, improving my mood and my outlook on life, and eventually even shifting my neuroscience research.

At the beginning of that class, the instructor, Patricia Moreno, the woman who created this class, told us that the term *intenSati*, comes from the combination of two words. *Inten* comes from the word *intention*. *Sati* is a Pali word (a language from India) that means "awareness or mindfulness." She told us that the goal of the practice of intenSati is to bring an awareness/mindfulness to our own intentions. She explained that we were going to be doing different movements from kickboxing, dance, yoga, and the martial arts, all the time shouting positive affirmations along with each move. I was not so sure about the shouting part, but Moreno was a completely riveting instructor so I stayed to experience this intriguing new class for myself.

That first class felt like an explosion of movements. Moreno started off showing us a simple yet energetic movement, such as alternating left and right punches. Once we got the movement down, she would then give us the affirmations that we would shout out along with that move. For example, with the punches, we said out loud, "I am strong now!" This move was called Strong. Each move had a specific name. We would do the first movement for a while, and then she would add a movement/affirmation combo, until we had strung fifteen or twenty different movements and affirmations together. Each particular set of affirmation/movement combos was written as a series with a specific message. The message was one of empowerment: The power of your mind, the power of positive action, the power of your body, and the power of positive thoughts over negative ones. It was a workout with a message.

Moreno told us that what we declare with our voices is powerful. And that when we start incorporating these powerful affirmations into our thoughts—that is, when we start to think and believe them—they become even more powerful still.

When we pushed our arms up in the air in an alternating fashion with our palms open and our fingers spread wide, we shouted, "Yes! Yes! Yes! Yes!"

When we punched up and down, we shouted, "I believe I will succeed!"

When we threw uppercut punches with alternating hands, we shouted, "I am inspired now!"

It was a workout for my body *and* my brain. Asking your brain to remember arm and foot work combinations, as well as affirmations to shout out, is asking your brain to work! There are also the words of the affirmations that the instructor is telling you, which you are also trying to remember—even before she says them. So your memory is also put to the test in an intenSati class.

Of course, I didn't appreciate all of the brain–body connections being made by intenSati after just one class. I was trying hard just to keep up and remember the movements—never mind remembering the affirmations at the same time! And it was hard. Shouting those affirmations while doing all the moves made you more out of breath than just doing the movements alone and upped the level of the workout considerably. I was also definitely a little shy at first about shouting out the affirmations. But there were plenty of regulars in class that night shouting with abandon, and once I managed to get the movements down, I got caught up in the fun and started shouting along with everyone else.

Have you heard people say that people won't remember what you say, only how you made them feel? I can't remember the exact affirmations that I said that night in class, but I do remember how I felt: totally empowered, energized, and enlivened—in a brand-new way. And I could not wait to come back for the next class.

HARNESSING THE POWER OF THE BRAIN–BODY CONNECTION

What was so different about this workout? Remember, I was already in good if not great shape by the time I wandered into this class. I was really starting to feel great about both my overall cardiovascular and muscular strength as well as the outside package after I lost the weight. I loved going to the gym and had already made it a regular part of my life. I was already feeling great and energized and was sure that my workouts helped me through those stress-filled years as I was applying for tenure, but intenSati brought something brand new into my life. I would not have been able to articulate it at first, but I now realize that this workout was so special because it brought the power of the brain–body connection to life for me more powerfully than I had ever felt it before.

The first thing I noticed was that I pushed myself during those workouts more than I had in any other class I was taking. Why? It was the power of those positive affirmations and actually speaking them out loud that seemed to flip a switch in me. It was the difference between doing a class and getting a good, sweat-inducing workout and really feeling strong because I was declaring I was strong or empowered or confident or a million other positive affirmations we used in that class. I was pushing myself even harder because I started to really believe I was strong. And I started to really feel that strength, embodying it not just during class but also long after class ended, when I went back into the real world.

But this is where the power of the brain–body connection comes into play. This connection refers to the idea that the body has a powerful influence on our brain functions and conversely that the brain has a powerful influence over how our bodies feel and work and heal. While I had been going to the gym for some time, and I definitely felt much more fit and energized and happy, I really started to appreciate the true power of the brain–body connection only with this new class. And the first thing I noticed was how strongly this workout (*body*) boosted my mood (*brain*).

From a neurobiological perspective, we know the most about the brain basis of mood from situations in which mood is altered—namely from the study of depression, one of the most common psychiatric conditions in first world countries like the United States. From studies of abnormal mood states, we know mood is determined by a widespread and interconnected group of brain structures together with interconnected levels of a set of well-studied neurotransmitters and growth factors. We talked about the role of the hippocampus in memory, and recent studies have shown that its normal functioning is also involved in mood. In addition, the amygdala, important for processing and responding to emotional stimuli, and the prefrontal cortex are both implicated in regulating our mood states. Furthermore, two other systems, which I describe in greater detail in later chapters—the autonomic nervous system including the hypothalamus (Chapter 7) and the reward circuit (Chapter 8)—are involved in regulating our mood. We also know that the appropriate levels of particular neurotransmitters are important for regulating mood. An influential theory of depression is that it is caused by a depletion of a category of neurotransmitters called monoamines. These include serotonin, whose low levels most of us associate with depression, but lowered levels of norepinephrine, another neurotransmitter, as well as dopamine are found in the brains of patients with depression. Therefore, the studies suggest that if you boost the levels of these neurotransmitters, you can boost mood.

Well, little did I know but I was getting a triple whammy of mood-boosting power with the intenSati workout. First, many studies have shown that not only does aerobic exercise improve measures of mood in subjects both with depression and without but that exercise boosts levels of the three key monoamines we know play a key role in mood: serotonin, noradrenaline, and dopamine. Besides these classic mood-associated neurotransmitters, exercise also increases levels of endorphins in the brain. *Endorphin* literally means "endogenous (made in the body) morphine." It is a kind of morphine that has the ability to dull pain and provide feelings of euphoria. Endorphins are secreted by the brain's pituitary gland into the blood, where they can affect cells throughout the brain that have specific receptors for them. Because endorphins are secreted into the bloodstream, they are categorized as a hormone; neurotransmitters, on the other hand, are released at synapses from the axon of the cells that synthesize them.

While most of us assume that endorphins are responsible for all or most of the high associated with some forms of exercise, the story is not as clear as all that. In fact, for many years there was a huge controversy in the neuroscience community (invisible to the popular press) over whether endorphins had anything to do at all with the so-called runner's high. This was because, while there was good evidence that the level of endorphins increased in the peripheral bloodstream (that is, the bloodstream that courses through the body), it was not clear if exercise changed the level of endorphins in the brain, which is where they had to be working to produce the runner's high. Only recently has a group in Germany provided evidence that running does activate the endorphin system in human brains and that the more profound the reported runner's high, the stronger the activation. So neuroscience shows that a range of different neurotransmitters associated with mood and/or euphoria are increased with exercise and are likely causing at least part of the party mood caused by exercise.

The second mood-boosting whammy from intenSati comes from the spoken affirmations that are such a prominent part of this workout. A relatively large body of psychology experiments has shown that self-affirmations like the ones we were shouting to the rooftops in class help buffer people from a whole variety of different stressors, including peer-based classroom stress, rumination associated with negative feedback, and stress associated with social evaluation. One recent study reported that positive self-affirmations significantly improved mood in people with high self-esteem. We don't know the brain and neurochemical changes associated with self-affirmations, but the behavioral evidence is quite clear that positive affirmations boost mood.

The third mood-boosting whammy of intenSati comes from the fact that during class, the physical moves that we perform are very strong and powerful; we are essentially in one power pose after another. The TED talk sensation Amy Cuddy, a social psychologist from Harvard, did a study in which she had some people pose in powerful positions with their arms behind their head and their feet up on a desk (the Obama pose) or with both hands leaning forward on a table in a pose of authority for just one minute and other people pose in nonpowerful positions, like sitting with the legs and arms crossed. The study showed that relative to the nonpowerful posers the power posers had increased levels of testosterone and decreased levels of the stress hormone cortisol in the bloodstream (after just one minute) as well as increased feelings of power and higher levels of risk tolerance. Indeed, recent studies in rodents confirm that exercise can increase testosterone levels in the blood, but other studies show that higher-intensity aerobic exercise *increases* circulating cortisol levels. These findings add to our knowledge of the powerful cocktail of brain and blood factors that can shift our mood after exercise.

This was also when I first started to see that while exercise is great for our bodies, when we make exercise both aerobic *and* mental—meaning you are fully engaged in the movement and/or feel passionately about it—we trigger another very powerful level of the mind-body connection. I call this *intentional exercise.* The key point is this: While the form of exercise I found in my gym was a fantastic example of intentional exercise, it is not the only one. I realized that you could make any workout intentional simply by bringing your own positive intentions, affirmations, or mantras to the class and focusing on those as you work out. Bring a positive affirmation like "I am sexy" or "I am graceful" to your next Zumba class. Choose a mantra like "I am strong" or "I am powerful" for a cardio/weight-training class or your next run. Adding your own personal affirmation or mantra to your favorite workout will do the same thing as what I experienced in the intenSati class. It will create the same positive feedback loop of affirmations and exercise, leading to good mood, leading to higher motivation, leading to higher levels of exercise, and leading to even better mood. You might need to play around with the kind of exercise you choose to optimize the effect. It has to be one you enjoy and that will allow you to get into the affirmations that really motivate you to do more. Try it out and see what works!

POWERFUL AFFIRMATIONS AND MANTRAS

People often tell me they can't think of their own affirmations. Here are some to help inspire positive intentions during your own workout:

I am inspired.
I am grateful.
I am sexy.
I am confident.
I am Wonder Woman strong.
I am Superman strong.
My body is healthy.
My brain is beautiful.
I throw away the old and embrace the new.
I expand my comfort zone every day.

I was feeling great, strong, happy, and motivated. I was making a powerful shift. Then I realized this shift in my mood was starting to change deeper things in me. While I thought I had high self-esteem, particularly around my work, I felt my exercise start to shift other parts of my self-esteem that had long been buried at the back of my closet, especially that part that was much more vulnerable in social situations. I looked better than I had in years and was in great shape, and my intentional exercise was starting to work on that negative part of my self-esteem in wonderful ways.

Yes, I had been a regular gym goer for several years by this point; but most of that time, I kept to myself at the gym. I came in, did my workouts with Carrie or in a class, and went home. With the positive effects of the intenSati class taking hold, I started making friends at the gym—maybe not lifelong BFFs, but friends nonetheless. It was quite a step for me because it was the first time I started making very exotic kinds of friends: non-scientists. It turns out, there was a whole world of people at the gym whom I had been standing next to in class for a couple of years; only with the influence of my intentional exercise, however, did I start to come out of my shell and begin talking to them. I met all sorts of new people, from stylists to actors to business-people to publicists. I found a new network of people to connect with, not just in the intenSati class but in all the classes I was taking. This new shift in my life started with fitness and weight, but it turns out that even more changes were a comin'.

Another noticeable shift came in a different area of my life: my teaching. Influenced by Marian Diamond and her wonderful teaching abilities, I have always loved and taken great pride in my teaching. I was as devoted to teaching as I was to setting up my lab in those early days as a new assistant professor at NYU, and I always brought great preparation, organization, and enthusiasm, if not a somewhat traditional approach, to the courses that I taught. Inspired by Diamond, I particularly loved teaching neuroanatomy and tried to bring a spark of interest to this rather daunting topic in the best way that I could. My efforts paid off because my teaching evaluations were always very strong.

But I can see now that all the shifts and changes that were happening in my personal life also affected my teaching in a similarly positive way. For example, one of the courses I teach to junior and senior neural science majors is called "Neurophysiology of Memory." I have students read a series of carefully chosen papers, and we then go through the history of our understanding of how the cells of the hippocampus and related brain areas fire when we are learning or remembering something. I was looking forward to the class, but I wanted to get playful with my teaching; inspired by a class at the gym, I knew exactly what I wanted to do.

I was taking dance classes that often left me frustrated because I just could not learn the choreography as well as other people in the class. Granted, many of the other students were former or aspiring dancers, so I should not have been so hard on myself, but I was still frustrated. I realized that part of my problem was that I was trying to explicitly memorize the steps that the teacher was showing me using my favorite brain structure, the hippocampus. But then it hit me; I knew motor learning used different brain areas that were non-declarative or unconscious. Think about it, you don't have detailed conscious awareness of all the muscles you use for your golf swing. You learn it with practice in an unconscious way.

That was it! That was the answer! I was just using the *wrong* part of my brain to learn the dance! Now all I had to do was figure out how to use my basal ganglia instead of my hippocampus, and Alvin Ailey, here I come! I did stop trying to memorize the dance moves and just tried to focus on moving my body in the way the teacher was showing and kind of going with the flow, which definitely improved my choreography learning curve noticeably. This was a real breakthrough in my own personal understanding of how memory works. I knew theoretically the differences between the two memory systems, but this was personal, real life, and the realization really helped me dance better in class. I had to try to focus on the right brain system and stop trying to overuse my not-so-great declarative memorization skills.

Even more important, this breakthrough inspired a new lecture session in my "Neurophysiology of Memory" class, which I could not wait to do. The idea was to teach students about the two different learning systems (and the neurophysiological studies that had been done on these two systems), contrasting declarative memory for facts and events with unconscious motor system–based learning (for example, learning the physical movements used to play the piano or tennis). I wanted to give the students the same experience of discovering the difference between the two different types of memory, so I had them learn a dance routine in class. It was a speed-dial version of the discovery I made after many, many dance classes, but I thought I could pull it off.

And I knew exactly the person I wanted to teach that class: my friend Erika Shannon, who had taught me intenSati during my teacher training and who was a sassy, vivacious, and playful dance teacher with a love of hip-hop. I wanted her to come to class and teach a hip-hop routine to my students to illustrate motor learning in the flesh. I remember I was so nervous when I asked her if she would do it. I think I was also a little nervous to do something so different in class. But I just *knew* the students would love it, and she graciously agreed to teach a routine for me.

I didn't tell the students what we would be doing but just asked them to wear something comfortable to move in to class. I had all the chairs cleared from the classroom, and they came in to find Shannon at the front, ready to take charge. I explained the premise of the class, and all the students were excited (if not a little trepidatious) to give hip-hop a whirl. The class was a smashing success and generated an enormous amount of discussion about the differences between the hippocampal learning system and the basal ganglia learning system and how one might differentiate the two.

That class happened because I had stepped out of my lab to see what else out there in the world might interest me. Not only that, I felt myself getting bolder, more creative, and more energized about what was possible. It was also an exercise in improving self-awareness for what I needed in my life. But I had more serious challenges directly ahead.

TAKING ON ANOTHER CHALLENGE: DATING IN THE CITY

I had started an avalanche of change in my life. Strength, fitness, weight, even new friends! But the question was, were all those positive affirmations enough to entice me to try my hand at dating? It had been so long since I dated anyone, I'd forgotten how! Or maybe I never learned very well in the first place. And my theory that men were just not that into me was still ringing loud and clear in my ears. What I really needed was an affirmation called "I will date now!" Despite the fact we never said that particular affirmation in class, buoyed by my new gym friendships, I took the big bold step to begin dating in New York City.

Bravely, I decided one day to give online dating a try. I created my profile and was immediately drowning in a sea of totally unappealing options. Were the interested men even telling the truth? What can you really learn from a picture and a few answers to questions like, What was the last book you read?

But I took the plunge. I ended up dating a cute guy in finance. We even went to the same gym, which felt like a positive recommendation in my book. In hindsight, I can see that his energy was really a mask for manic tendencies. While he was great fun to be with in the manic stage, he eventually disappeared, never to be heard from again.

Next!

I put the online dating sites on hold after that experience and moved on. I decided I needed a few moments to catch my breath and that I might have to bring in the big guns. I had read about matchmakers and how they get to know you and prescreen dates for you, choosing from only the men they believe will be a great match. In other words, the matchmaker and not I would be doing the equivalent of sorting through all the profiles of men who take a picture of themselves holding a camera in front of their bathroom mirror with their frayed toothbrush and razor in the shot for all to see. Yes, that was exactly what I needed.

So I googled "Matchmaker NYC" and found one that sounded reasonable, and made an appointment. I found myself in a sterile corporate suite that she clearly rented out for just these kinds of interviews. She seemed to have a wide selection of eligible, highly educated, intelligent men and a reasonable fee so I decided to give her a try.

My first date was with a guy in the banking industry but currently "between jobs." He suggested we meet for a drink at a bar in Bryant Park and we had a very nice chat in which he impressed me with the books he had read and the people that he met hanging out at this very bar (Malcom Gladwell of *Blink* fame, for one). This seemed like a good start, and our second date was even more promising because he asked me to come dancing with him at one of the Midsummer Night Swing parties, a yearly event at Lincoln Center. You go an hour early and get a big group dance lesson and then you stay for the live band that plays swing, Lindy hop, or salsa, depending on the week. I was thrilled! I'd been wanting to go to that event for years but had never had a partner to accompany me. I couldn't wait.

Yet, not everything was Prince Charming perfect. This guy had an annoying habit of pulling out a little comb from his pocket and combing his hair for apparently no reason at all. A nervous trait maybe? And then he kept talking about the fact that he had a car. He talked about how this car was so convenient to get away from all the chaos in the city and how lucky he was to have it.

Hence the nickname I gave him: "Car Boy."

One day, Car Boy called to invite me on a "car-centric" date. He proposed that we take a drive in his car down to Philly to visit the Barnes museum. I had never been to see this formerly private collection. I love museum visits so I accepted. We were set to meet on a sunny Saturday morning.

I should have realized immediately that something was wrong when a few days before he asked if I could meet him at his place on the West Side at about 10:30 on that Saturday morning. Now, wait a minute. Car Boy is asking me to take the subway across and up town on the morning of our date instead of using his fancy car to come and pick me up? Most women, I realized only now, would have immediately canceled or at least demanded to be picked up in his *car*, but no, I said to myself, maybe it's just a kind of New York thing and I made my way on the subway too early on a Saturday morning all the way across town to meet Car Boy and his fancy car.

When I finally arrived at his apartment, Car Boy buzzed me in, and we made our way down to the garage. I must admit the car was beautiful. It was a big Jeep or Range Rover kind of model. It just seemed way too big for a single person to have in the city, but very nice nonetheless. I perked up a little and sat myself down in his big fancy car for the ride down to Philly.

The ride was uneventful and the museum was wonderful, but the more time I spent with him, the more annoyed I got with myself for actually agreeing to tromp all the way over to his house instead of insisting that he pick me up. The more I thought about my mistake, the more I wanted to get out of the car and away from this man. By the end of the excruciatingly long car ride back, where I had nothing to do but sit there in his fancy car and couldn't think of anything I wanted to talk to him about, I could not get away from Car Boy fast enough. You won't be surprised to learn that he didn't even offer to drive me home either. He did agree (just a tad begrudgingly) to stop the car so I could catch the subway home. If I had not asked, I think he would have just let me off at his place to make the long cross-town trek on my own.

Next!

I was willing to give the matchmaker another chance; besides I had already paid her for five introductions so I might as well see if she could come up with someone better than Car Boy.

Next up: Cabin Boy.

Cabin Boy specialized in the high-tech field. On our first date, we had a lot of fun. The food at the restaurant he had chosen was nothing special, but he had a lot of interesting stories about all the high-tech gadgets he was involved in developing. He also told me about his prized cabin or "dacha," as he liked to call it. Apparently his magical getaway from the stressful city was in the middle of nowhere, yet easily accessible from New York, quiet, pristine, and secluded. A perfect hidden treasure.

I was intrigued.

After a lot of flirtatious e-mail back and forth, we had a few more dinner dates, during which he continued to talk about his cabin among the high-tech things. He was interesting and smart, if a little distant, and I was enjoying going out with him. I also found myself becoming a little obsessed with his cabin. Would I see it? How far away was it? How often did he go? Was there a bearskin rug on the floor? Was I more intrigued with the cabin than with Cabin Boy himself? I was asking myself all but the last question.

One Thursday afternoon he e-mailed to see if I wanted to drive out to his cabin the next day for a drink. The *next day*? For a drink? What kind of weird last-minute invitation was that? He didn't tell me how long it would take to get there, if there would be a possibility of food after the drink, or how long the drink invitation would last. I declined, yet part of me really wanted to see the magical cabin in the woods.

Well, that mythical cabin was never meant to be. About a week later, when I was at a conference in Colorado, Cabin Boy called to tell me he was seeing someone else and that he had decided to "take the high road" and break it off with me.

All I could think of in response was, You mean rather than taking the low road and keep dating me?

Next!

After that episode, I took a break from the car boys and the cabin boys of my dating world and retreated back to the gym where

I thought I belonged. I had not given up hope, I just needed to re-group and reconsider my strategies. It was the first time out of the gate for me. I think I needed more practice or better luck or both. But the point was I *did* it! I was slowly resuscitating my sickly social life into something interesting. Not perfect yet, but getting there. Things were looking up. I was getting curious about lots of things that I had not experienced before. Dating was just a kind of hobby or curiosity for me at that point. I still very much believed in my old theory that men were simply not that into me. But I was about to turn my curiosity on another part of myself. I started to wonder exactly what was the origin of my sky-high mood and energy. In other words I started to get super curious about how my workouts were really affecting my brain.

HOW TO MAKE YOUR WORKOUT INTENTIONAL

The unique thing about intenSati is that it pairs positive spoken affirmations with aerobic workout moves. The affirmations not only increase the cardiorespiratory level of the workout but add an intentional component to the workout. But any workout can be intentional. All you have to do is add a powerful or uplifting or fun mantra or affirmation to your own favorite workout. For example, while jogging you could be chanting, "I am strong now!" in time to your stride. While biking, you could say, "Today, I am inspired!"

Saying the affirmations out loud is great because it strengthens the declaration. But sometimes you can't shout affirmations out loud—for example, if you are in a big class at the gym with lots of other people. In that case, you can just choose your favorite mantra or affirmation and either just have it run through your mind or just declare it softly to yourself. For example, kickboxing class or spin class or cardio boot camp classes would all be great possibilities to include an affirmation like "I am powerful" or "I have no fear of mistakes" along with your workout.

Another great example is if you are somewhere out in nature on a hike or a bike ride all alone with your thoughts. Or try sharing the experience with someone else; then you can say your affirmations together and feel the powerful effect of shared intentional exercise.

TAKE-AWAYS: THE BRAIN–BODY CONNECTION

- The brain–body connection is the idea that your brain, including your thoughts, can affect your body (for example, thinking positive thoughts about a healing injury or recovery from the flu can speed the process), and conversely changes in your body (increased or decreased movement, for example) can affect your brain.
- Intentional exercise happens when you make exercise both aerobic (movement) and mental (with affirmations or mantras). You are fully engaged in the movement and trigger a heightened awareness of the brain–body connection.
- Intentional exercise can boost your mood even more than just exercise alone.
- You can make any exercise intentional by adding affirmations or mantras to your workout.
- Other brain areas involved in the regulation of mood include the hippocampus, amygdala, autonomic nervous system, hypothalamus, and the reward system.
- Exercise enhances mood by increasing brain levels of the neurotransmitters serotonin, norepinephrine, and dopamine as well as increasing the levels of the neurohormone endorphin.
- Positive affirmations have been shown to boost mood but the neurobiology underlying this behavioral change is not yet understood.
- Just one minute of power poses decreases the stress hormone cortisol, increases testosterone, and results in people performing better in interview situations, suggesting that power poses should be used to prepare for important talks, presentations, and interviews.

BRAIN HACKS: FINDING THE EXERCISE THAT'S RIGHT FOR YOU

Are you currently in search of an exercise that will motivate you to get up off the couch and move your body on a regular basis? An exercise like intenSati really upped my dedication and enjoyment of exercise. The trick is to search out that particular form of exercise that you enjoy the most. There may be lots that you don't enjoy, but find the one that really makes you feel great. Here are some avenues to explore:

• If you love the outdoors and that wonderful sensory experience of nature, then definitely choose an activity like hiking, walking, or biking outside.
• You can help enhance any workout with the music that makes your toes tap. Spend some time exploring an online music store, virtual radio app, or a music video station on YouTube and find those songs for yourself. I know for me a great song can make me start moving when I thought I was done for the day.
• If you love working out with others, find some friends to exercise with or make new workout buddies at a gym.
• For me, a great instructor can make me work out much harder and have much more fun than I would have on my own. See if you can find an instructor like that and take his or her class.
• If you already like to do things like dance or ski or hike, then incorporate those into your regular workout schedule.

And finally, just keep this in mind: If you are learning a new exercise, don't expect it to give you that endorphin high the very first time you do it. You need to develop a certain level of expertise in an activity before you can really start feeling that exercise high. So if you don't feel it at first, but enjoy the exercise, stick with it and wait until you develop more skill. The high will come. Just trust your gut.

5

THE BIRTH OF AN IDEA

How Does Exercise *Really* Affect the Brain?

A s I neared the deadline for yet another science grant application to NIH (something I spent a great deal of time on as a science faculty member), I was becoming aware that my writing was going well—unusually well. My daily writing sessions were much more productive and even *enjoyable* compared to the stress-filled sessions of the past. Whereas it usually took me a week to write just one section of a grant application, I was now drafting more efficiently, fine-tuning more quickly, and enjoying the process a whole lot more. My attention was more focused and my thinking more clear. I made deeper, more substantive connections between my ideas and was doing so far sooner in the process than usual. This is when I began to suspect that there was a relationship between the regularity of my workouts and my supercharged grant-writing sessions. The writing just went more smoothly during a week when I exercised more than three or four times, compared to weeks when I slacked off and went to the gym only once or twice.

What was going on here? I realized that without knowing it, I had just conducted an experiment on myself! I varied my exercise regime (some weeks four or five exercise sessions and other weeks only one or two) and found that only with a higher frequency of exercise did I notice enhanced attention and the ability to make new and better thought "connections" in my writing. While focused attention is known to depend on the prefrontal cortex, the ability to make new connections or associations is thought to depend on the hippocampus.

This was fascinating! I knew there had been lots of progress in our understanding of how exercise could affect brain function, but I had not kept up with that literature—too busy getting tenure, I guess. So when I noticed these changes in myself, I dove into that neuroscience literature to see what was new.

What I found was an active and growing body of research that was deep into identifying all the different ways aerobic exercise affects brain function. These studies documented a range of anatomical, physiological, neurochemical, and behavioral changes associated with increased aerobic exercise. But the biggest surprise I got when exploring this literature was that one of the founders of this whole line of research was a scientist with whom I was very familiar: Marian Diamond.

It seemed like a sign.

It turns out that our understanding of the effects of exercise on brain function had its origins in the original studies that Diamond did on brain plasticity and the effects of raising rats in enriched environments on brain function. As I mentioned in Chapter 1, those early studies showed all sorts of brain changes when rats were raised in the enriched environments: The animals developed a thicker cortex because of the more extensive branching of the dendrites, more blood vessels, and higher levels of particular neurotransmitters like acetylcholine as well as increased levels of growth factors like brain-derived neurotrophic factor (BDNF). Acetylcholine was the very first neurotransmitter ever discovered, and the brain cells that use it send signals throughout the cortex as well as to the hippocampus and amygdala. Acetylcholine is an important modulator of learning and memory, and studies show that drugs that interfere with the action of acetylcholine result in memory impairments in both animals and humans.

BDNF is a growth factor in the brain that supports the survival and growth of neurons during brain development as well as synaptic plasticity and learning in adulthood. Not only that, but a truly exciting finding was reported in the 1990s, when researchers in California demonstrated that rats raised in an enriched environment had more new neurons in their brains than other rats. This process is called neurogenesis. While lots of new brain cells are born during our early development period (from infancy through adolescence), there are only two places in the brain where new brain cells are created in the adult brain. One is the olfactory bulb, the part of the brain important for sensing and processing smells (see Chapter 1), and the second is my old friend the hippocampus. But even more significant is that new brain cells are formed on a regular basis in the hippocampus of adult rats. The enriched environment also was linked to a higher number of hippocampal brain cells (but not in the olfactory bulb). Other studies showed that rats raised in enriched environments and that had more new hippocampal cells also performed better on a range of different learning and memory tasks, suggesting that all these new neurons were helping the rats learn and remember better.

But then neuroscientists began to wonder what it was about the enriched environment that was causing all these striking brain changes. Was it the toys? Was it the gaggle of other rats to play with? Maybe it was all that running around that the rats did in the Disney World–like environment. When scientists tested these factors systematically, they discovered that one contributed to the majority of the brain changes seen with an enriched environment: exercise. They found that all they had to do was give a rat access to a running wheel, and they would see most of the brain changes they observed in rats that were raised in the enriched environment. Indeed, this line of research in rodents has shown us how exercise affects the brain at the molecular, cellular, brain circuit, and behavioral levels.

We now know that exercise alone can actually double the rate of neurogenesis in the hippocampus in rodents by increasing the total number of new neurons that are born, enhancing their survival (many of the new cells die) as well as speeding their maturation into fully functioning adult brain cells. These new neurons are not born just anywhere in the hippocampus but in only one specific subregion, called the dentate gyrus. When I read this, I felt like going to the gym and working out even harder! Exercise in rodents also increases the number of dendritic spines on the neurons in the dentate gyrus; these spines are budlike appendages on the neurons' dendrites, the branchlike structures where neurons receive information. Exercise also increases the total length, complexity, and spine density of the dendrites. Thus it is not surprising that the total volume of the dentate gyrus also increases with exercise. Spine density in other regions of the hippocampus and the adjacent entorhinal cortex (which I studied as a graduate student) is also increased significantly by exercise. Spines are where the axon of one neuron contacts the dendrites of the next neurons; the more spines on a neuron, the more communication is happening. Another robust change researchers confirmed with exercise alone was the growth of new blood vessels throughout the brain (including in the hippocampus), which is called angiogenesis.

The physiological properties of the rodent hippocampus also change after exercise. This physiological phenomenon is called long-term potentiation (LTP), which is a long-lasting change in the electrical response between two groups of neurons. We study this change by stimulating the connections between the two groups of cells in the hippocampus with an electrical current. If you stimulate one of the pathways within the hippocampus with multiple fast bursts of electric current this will increase the response you get from a weak electrical stimulus to that pathway compared to the same weak stimulus given to the pathway before the burst of current. LTP is widely considered to be a major cellular mechanism for learning and memory. LTP is enhanced in the brains of rats that have been exposed to exercise. One key factor that might be contributing to these effects is the increase in BDNF because we know that BDNF can also enhance LTP. But BDNF is not the only factor that increases with exercise. As I mentioned in Chapter 4, in addition to all these anatomical and physiological changes, exercise also increases the brain levels of serotonin, norepinephrine, dopamine, and endorphins.

Given that exercise increases the number of new brain cells, beefs up the size of the cells in the hippocampus, enhances BDNF and LTP, and increases the level of transmitters and growth factors floating around the brain, the next logical question is, Does exercise enhance the function of the hippocampus? Do rats that exercise more remember better? Indeed, many studies have shown that rats that have either undergone environmental enrichment or exercise alone perform better on a wide range of memory tasks that we know depend on the hippocampus. These include spatial maze tasks, memory-delay tasks, recognition-memory tasks, and a range of memory-encoding tasks. In these latter tasks, rats are asked to differentiate between similar items during a memory task. Memory encoding is a function that neuroscientists believe specifically depends on the dentate gyrus, the subdivision in the hippocampus where all those new neurons are born. More generally, as we first learned from H.M., the hippocampus is known to be important for both learning (acquiring new information) and retaining that information. You can't learn information and store it as a long-term memory without a hippocampus, but information learned before an injury to the hippocampus remains intact. While we know the hippocampus is critical for the formation of our long-term declarative memories for facts and events, we still don't know exactly how it does this amazing feat. That's the topic of an enormous amount of current neuroscience research. I believe that things like LTP are involved and levels of BDNF also help the process, and we also know a lot of the molecular pathways involved. However, we are still filling in all the pieces of our understanding of exactly how the hippocampus in general and the dentate gyrus in particular (the part of the hippocampus where those new neurons are born) work to form all the new memories that we make every day.

On the other hand, I don't *need* to know exactly how it works to appreciate the fact that memory performance is better with exercise. Now *this* was exciting news! If rats that run have better memories, then people who exercise should also have better memories, right? I certainly noticed an improvement in my own ability to make connections or associations in memory I know depends on the hippocampus, and studies in rodents suggested that I was on the right track.

CRAFTING MY OWN BRAIN HACK

Then I had an idea.

You know when you come up with an idea that you are so excited about you simply can't wait to make it happen? That's how I felt. The idea started with my desire to get up to speed on this exciting new neuroscience literature on the effects of exercise on brain function. As any teacher worth her salt knows, the very best way to become familiar with and understand a specific area of research is to teach a new course on it. So I decided to teach a new neural science elective course on the effects of exercise on brain function. I was feeling especially energetic and creative from all the exercise I was doing and was inspired to develop the new class because of my own exercise experience. What if I brought exercise into the class and not only told the students about the neuroscience underlying the effects of exercise on brain function but actually let them *experience* the effects of exercise themselves? Of course the form of exercise I wanted to use was intenSati—my exercise of choice at the time. I just knew that adding exercise to the classroom would bring the course to a whole new level and send motivation for learning the material through the roof.

That's right: I was going to bring aerobic exercise *into the university classroom* for the first time ever, so students could literally experience the positive effects of exercise as they were learning about what exercise was doing to their brains. With this idea, my "Can Exercise Change Your Brain?" class was born. It would be uniting two of my greatest joys—teaching and exercising—in a unique way. I was already imagining the format of the course: We would start out each class with an hour of intenSati and finish with a ninety-minute lecture/discussion starting with the history of the study of the effects of exercise on brain function. I would end the semester with the studies in humans, describing what we currently know about how exercise affects cognition.

I was beyond excited!

But I knew I was getting a little bit ahead of myself, and I needed to figure out how I was going to get an intenSati instructor into my classroom every week to teach in my class. The problem was that there was no money available to hire an instructor to teach the exercise part of the class. As a professor of neural science at NYU, I am expected to teach all aspects of the courses I design. Well, the obvious solution in my exercise-soaked mind was to learn how to teach intenSati myself.

To tell the truth, a secret part of me was just looking for a good excuse to learn how to teach intenSati. I remember the day I became aware of my desire to teach this class for the first time. I was standing in the studio waiting for an intenSati class to begin, and I started chatting with Pattie, a very sweet woman whom I had become friendly with in class. She mentioned casually that she had taken the teacher-training class and how much she loved it. My ears perked up immediately! I thought you had to be some kind of fitness goddess or triathlete to take a teacher-training class. But Pattie was just like me—a regular person who happened to be a devoted student in the class. I was intrigued, and I was more than a little envious that she had learned the secrets to teaching the class that had kept us both coming back time and time again. So, when the possibility that I needed to be the one to teach the intenSati part of the class surfaced, I jumped at the chance.

I think the other reason I was so excited to learn how to teach the class was that the idea woke the inner Broadway diva in me, who had been snoozing for a very long time. Granted, I would not be belting out "Defying Gravity" from *Wicked,* or "Let It Go" from *Frozen,* but I would be shouting affirmations with the beat of the music and leading the whole class in the mini–dance routines that made up intenSati. Maybe this was my chance to bring a little bit of Broadway onto my own private stage: the academic classroom.

Yes, there were lots of reasons motivating me to jump right into teacher training. But there were other reasons that held me back. For one, the fear of becoming a big fat Broadway-style flop. I was a great student in class, where I could follow all the instructions with ease, but could I actually be the one to *give* the instructions and not mess up? Second was the possibility of ridicule by my faculty colleagues, most of whom generally had a much more conservative style of teaching than I did. There were no other courses taught either in my department or at the whole university that I knew of that resembled this course at all. It was brand-new territory.

I would have to put myself out there in a way I never had before— not just before those who would take my class—and in spandex— but for all my faculty colleagues. They would think I was crazy. I knew it. And because I was going to be teaching in our main neural science classroom, the whole class and I would be front and center for anyone walking past to see us jumping, punching, and kicking (not to mention the heart-pounding music in the background that was bound to attract attention). Just thinking about the various ways that my plan could unfold filled me with a mixture of fear and excitement. And that's when I knew this wasn't just a silly fantasy; I had to do it. So before I could change my mind, I signed up for the intenSati teacher-training class at the gym and wrote up and submitted my class syllabus. There was no turning back now.

Despite the fact that I had never taught an exercise class before, I jumped right in. The five days of intenSati teacher training in New York translated into eight hours a day at the gym, studying the physical movements associated with the class but also learning about the ideas behind the practice, including aspects of positive psychology, personal coaching, and how to motivate people more broadly.

THE EXERCISE PLOT THICKENS

Soon after I first came up for the idea for the class, I realized that it had the makings of something more than just a unique new undergraduate experience. It had the makings of a full-on research study with human subjects who happened to be my students.

Given the exciting research on the effects of exercise in rodents (and the enthusiastic articles in the popular press coming out all the time), one would expect a rich literature on the effects of exercise in humans. But it is relatively modest and biased either toward studies of the effects of exercise in the elderly population (typically defined as sixty-five years old and older) or the effects of exercise on school-age children. There is very little information on the effects of exercise on healthy adults, like me. In other words, there is an enormous number of important questions left to answer in the human population. The studies focused on the elderly consistently show that the amount of average physical exercise reported over the lifetime is strongly correlated to your brain health as you age, with the highest levels of overall exercise correlated with the best brain health. For example, one representative study surveyed 1,740 participants over the age of sixty-five who did not have cognitive impairments and asked about their exercise frequency, cognitive function, physical function, and depression levels. Scientists then followed up with these same people six years later asking how many of subjects developed dementia and/or Alzheimer's disease. They then went back and looked at how much the people who did or did not develop dementia or Alzheimer's disease exercised over their lifetime. The big take home was that people who reported exercising three times a week or more had a 32 percent reduced risk of developing dementia. Anyone with family or friends with dementia can appreciate that a 32 percent risk reduction is big . . . really big. It's these findings and others like it that have encouraged researchers to try to more fully understand the cognitive effects of exercise, and whether those effects might be maximized. Similarly, studies in school-age children have shown that aerobic fitness has a small but positive relation to academic achievement, and body mass index (BMI) has a negative relation to academic achievement (that is, high BMI is associated with lower levels of academic achievement).

But these studies alone are not the final word on this topic. Not by a long shot. These kinds of studies are called correlational because they come to a conclusion by comparing/correlating self-reported levels of exercise (researchers have no control over the amount or quality of exercise and cannot judge the accuracy of the self-reports) with the subject's current state of brain health. While such studies suggest the possibility that the level of physical activity over a lifetime has an effect on brain health and dementia as one ages, there are also other possible explanations that cannot be ruled out. For example, maybe all the people who exercised more came from a higher socioeconomic status. Or maybe all the people who exercised more were just overall healthier, with better and stronger hearts. This would suggest that socioeconomic status or overall health defines how healthy and free of dementia your brain is when you get older and not amount of exercise. For these reasons, the conclusions from correlational studies, while informative, are far from definitive. For example, in addition to the studies I mentioned that link higher levels of exercise to lower incidences of dementia, other more direct studies have also linked increased levels of exercise to better learning and memory performance. Again, it's a promising direction, but still not conclusive.

So what's more powerful than an observational study? The gold standard is what's called an interventional study. Another term for this same style of study is randomized controlled study. For this kind of study, you take a group of subjects and randomly assign them to either an exercise group or a control group that does not exercise. In this way the experimenter has direct control over the manipulation being done. Then you are able to compare the performance of the exercise group to that of the control group to determine if exercise has any significant benefit relative to the controls based on the factors that you've determined. Few of these gold standard types of exercise studies have been done in the elderly, but such studies have shown that an exercise intervention for several months to a year results in sharper attention, faster response times, and improved visuospatial functions (cognitive functions that require manipulating visual and spatial information in memory). Indeed, the largest and most consistent effect in the elderly seems to be on attention or the ability to focus or concentrate on discrete aspects of information while ignoring other perceivable information. Improved attention is an effect that I noticed clearly in myself with increased exercise. Similarly, one randomized controlled study showed an increase in the size of the hippocampus in elderly subjects who exercised for a year, which is consistent with findings in rodents. Another study reported significant increases in vasculature in the hippocampus after a three-month exercise regime, associated with subtle improvement on a memory task.

RANDOMIZED CONTROLLED STUDY DESIGN

The gold standard for how to test the effect of an intervention like meditation or exercise on a group of people is to use what is called a randomized controlled study design. In these kinds of studies, people deemed appropriate for the study (the right age, background, and health status) are randomly assigned to either a test group or a control group. The test group is assigned to do the manipulation you are interested in studying, like exercise. The control group is assigned to do something that has all the elements of the thing you are doing in the test group but not the key element you think makes the most difference. So a good control group assignment for exercise would be slow walking, for example. Another important component is the idea that you are going to test. In this kind of study, you might test the idea that aerobic exercise improves memory function relative to the control manipulation of slow walking. To examine this idea, you will test both groups on their memory performance before and after either aerobic exercise or walking. Then you can ask if the aerobic exercise group improved their memory test scores significantly more than the walking group. If so, this would be a strong indication that aerobic exercise improves memory. The power of putting your subjects in groups randomly is that you can rule out that there were any major differences between the groups before the study started because you not only randomly assigned each person to a group but you have test scores before the intervention to show that there were no differences on the memory tests. This is the gold standard for experimental study design.

Fewer of these valuable randomized controlled studies have been done because they are much harder to carry out and generally more expensive than correlational studies. But what you get out of the randomized controlled studies that you can't get from the correlational studies is the prescriptive aspects of exercise. With randomized controlled studies you can say, "We showed that X amount of Y kind of exercise improved Z brain function." This is exactly the kind of information we need from research with humans. We don't know what kind of exercise works best, what duration and activity level is best, or if men and women have different optimal exercise regimes for optimal brain health. And a big one we don't know: What exactly is exercise doing to the elderly brain?

CAN EXERCISE MAKE ME SMARTER?

In contrast to all the work in the elderly, much less is known about the effects of exercise in healthy young adults. This is because it is generally thought that healthy young adults are at the height of their brainpower and therefore have little room to improve. In contrast, the cognitive decline in aging is a normal occurrence, so there is a larger window of possible improvement in this group relative to younger adults. But if there is little or no effect of exercise on brain functions in young, or at least youngish, adults, then why did I notice such a striking effect on my grant writing after my self-imposed exercise regime? The bottom line is that there were so few studies on the adult, nonelderly population that no strong conclusions could be made. I wanted to change that.

While neuroscientists had not made a lot of headway in examining the effects of exercise in healthy young adults, the possibility of creating a magic pill to make us smarter has fascinated us for ages. Books like the classic *Mrs. Frisby and the Rats of NIMH*

(1971) and short stories like "Flowers for Algernon" (1959) have explored the possibility of making both men and rats smarter. In the movie *Limitless* (2011), Eddie Morra (played by Bradley Cooper) is a down-on-his-luck loser with bad hair and even worse clothes until he stumbles on an illegal but powerful pill that enhances cognitive power. After taking it, he immediately makes a ton of money on the stock market, gets a good haircut and a fancy new set of clothes, and is livin' large. That is, until the side effects of his smart pills set in. In the end, he becomes so smart that he finds a way to engineer the magic pill so he suffers none of the side effects but retains all of the cognitive benefits: He runs for Senate and learns to speak fluent Chinese. Amazing! In *Rise of the Planet of the Apes* (2011), Will Rodman (played by James Franco) develops a new substance (this one came in the form of an inhalable gas) that improves cognitive function in Alzheimer's disease patients by repairing the damage that the disease causes. The drug is too late to save his father, but his young pet ape Caesar gets a whiff and suddenly learns how to speak and take over an entire city! Now that's one powerful drug. Clearly, a smart pill is an object of endless fascination.

But those are all works of pure fiction. Back in the real world, while exercise will probably not work as well as Eddie Morra's pill or Will Rodman's gas, I saw evidence in myself that intentional exercise could have a clear and noticeable effect on a range of brain functions that I used every day. And my review of the research showed that there was little known about the effects of exercise in young adults. Now I had the perfect opportunity to address this question with my "Can Exercise Change Your Brain?" class. My students would be exercising once a week for an hour, for the fourteen weeks of the semester. I was missing only two elements to turn this class into a real research study. The first was the ability to test the students' learning memory and attention abilities both at the beginning and at the end of the semester. The second was a control neuroscience class that met for the same number of hours but that did not exercise, so I could also test those students at the beginning and at the end of the semester. While this study had many elements of a gold standard interventional study, it was not perfect. A true gold standard study would have randomly assigned students into either an exercise elective class or a nonexercising elective class with the same instructor for both. Instead, I gave exercise to the students who wanted to take the exercise class and compared them to students who did not sign up for the class and were being taught by a different teacher. While this was better than an observational study, we did not have the gold standard randomized controlled design because we couldn't randomly assign students with no consideration for their own personal choice of class. However, this somewhat less than optimal design would do just fine. The other factor that I had to take into consideration was that the class met only once a week, so I would get only a once-a-week exercise boost in these students. Had I designed this study outside the classroom, I would have wanted the students to exercise something like three times a week. But, in the end, this was just a preliminary classroom experiment. I was doing it to engage the students in a new way in the experimental process, and we would discuss all the ways in which the experiment was not optimal; I'd make that part of the learning experience. In fact, with the relatively low number of students in the classes (relative to major clinical studies), I was stacking the decks against myself to see any effect. But then again it meant that if we *did* see any effect—any effect at all—that would be very exciting.

I also made the explicit decision to use intenSati as the exercise in my experiment because of its integration of conscious intention. I chose it instead of using a more "pure" form of aerobic exercise, like treadmill running or pure aerobics or kickboxing, because my goal in this study was to examine the effect that I saw in myself. I wanted to determine if this particular approach held up in a study; if it did, then in future studies I could separate out the individual effects of the affirmations and exercise. I also thought that intenSati would engage and motivate students and be a better fit for my classroom-based study than having the students do a hard-core treadmill run every morning before class.

It turns out that a control neuroscience class that didn't exercise was easy to find in my department. The expertise in human testing was little trickier. Because I did not have much experience testing humans, I sought out help in this arena. But just as the ideas were swirling around in my mind, I happened to run into a colleague who became my collaborator on this project. Scott Small, a neuroscientist and neurologist from Columbia University, happened to walk by as I was sitting on the steps at the main entrance of a convention center in Washington, D.C., resting my feet after a morning of walking around a big neuroscience gathering. We started chatting, and I told him of my new interest in exercise and the brain and my plans to teach the new class. Turns out he and his colleague Adam Brickman, another neuroscientist and neurologist at Columbia, were also studying the effects of exercise on cognition in humans! They were interested in collaborating and getting more data from healthy young adults to support their first findings. I was thrilled at the prospect, and when we got back to New York we starting planning our new exercise research study based on my class. Sometimes you make the best science connections just sitting on some steps, resting your feet.

EXERCISE AND NEUROGENESIS

What did we think exercise would actually do to the brains of my healthy, young, high-functioning neural science majors at NYU?

The answer to this question comes from the studies on the effects of wheel running on hippocampal neurogenesis in rats, which I described earlier.

This whole line of research had a really exciting and controversial history dating back to the 1960s. It took a lot of work to convince people that neurogenesis could actually take place in the adult brain. For a very long time, it was believed that once you reached adulthood, no new neurons could be born in your brain. This idea was established and widely held in the neuroscience community well into the 1990s, despite the fact that two decades earlier a couple of researchers from Boston University published the first evidence that new brain cells could be generated in the adult brains of rats.

Unfortunately, by that point, the idea that the adult brain was fixed was so strongly entrenched in the minds of scientists that this early study did not make much of an impact. After about twenty years, those early researchers were finally vindicated in a series of studies using more modern (and more convincing) approaches to show definitively that shiny new neurons were born in adulthood in both the hippocampus and the olfactory bulb. Not only that, but in 1998 an international team from Sweden and the United States provided the first direct evidence that neurogenesis was occurring in the adult human hippocampus. They did this in a very clever way. In rodents, you can confirm the presence of newly born neurons only by first injecting the brain with bromodeoxyuridine (BrdU), then sacrificing the animals and examining their brains. Brain cells that incorporate this chemical have recently divided (that is, they were recently "born"), and many such cells were seen in the adult rat hippocampi. The researchers knew that BrdU is commonly used to test for cell growth/division in tumors in cancer patients, so the research teams went out and obtained permission from cancer patients who had been injected with BrdU to examine their brains after they died. From the brains they were able to examine, the researchers found that, just as in rats, these adult patients had BrdU-stained cells in their hippocampi (remember that we all have two hippocampi, one on each side of the brain). This confirmed that in adult humans, just like in adult rats, new hippocampal cells are born.

This is the question we would focus on in my "Can Exercise Change Your Brain?" class. That is, would the increased aerobic exercise from the intenSati class I would teach enhance neurogenesis in the hippocampi of my students and, as a consequence, improve memory function? We were seeing indications of improved cognitive functions in the elderly after exercise, even though neurogenesis decreases as we age. My class would test the idea that young NYU students would have heightened levels of neurogenesis and, therefore, have a good possibility of benefiting from increased aerobic exercise. We would not be looking directly at neurogenesis, but would measure it indirectly based on the students' performance on cognitive tasks that depended on the brain areas where the new brain cells were being born. This is the idea we were going to test.

fMRI

fMRI stands for "functional magnetic resonance imaging." Like standard MRI (see "MRI" on page 54), fMRI is also done with a big magnet but it detects the change in blood flow related to the energy used by the brain. We know that when a brain area is active, blood flow to that region increases and, in addition, there is a change from oxygenated to deoxygenated blood in highly active areas (the brain is the single highest user of oxygen in the body). fMRI provides an indirect measure of activity in specific brain areas by detecting changes in blood flow and oxygenation levels and is the most common tool used to measure brain activity in people.

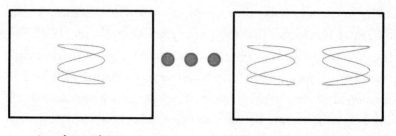

Look at this   **Which one did you see?**

A memory-encoding task.

This is where my Columbia University colleagues Scott Small and Adam Brickman came in. They had used a human brain imaging technique similar to functional magnetic resonance imaging (fMRI) to study active brain areas as subjects performed various tasks.

In one particular study, they asked subjects to commit a complex figure to memory (called memory encoding) and then identify that same complex image relative to similar complex images. Here is the protocol: As subjects performed this challenging memory-encoding task, Brickman and Small saw that the same specific subarea of the hippocampus where all the new neurons are born lit up like fireworks. The fact that this region was very active during the task, suggests the possibility that if we were able to use exercise to rev up this same brain area and create more new brain cells, we might see even better performance on this task. So the idea, or hypothesis, that we tested in this study was, Would an increase in aerobic exercise improve performance on this memory-encoding task?

And what does it mean that exercise can improve memory encoding? My lab and others are testing the idea that increased aerobic exercise enhances neurogenesis in the hippocampus, and those new hippocampal cells (because we know they are more excitable than the cells that have been there for a long time) improve our ability to encode, or lay down, new long-term memories. In particular, there is evidence that these newly born hippocampal cells help differentiate between incoming stimuli that have similar characteristics. For example, when I try to remember whether Julia or Pam came up after class to ask a question, it's the new hippocampal cells that help differentiate between the two, who both happen to have medium-length brown hair. What are the implications? That long-term exercise can increase the number of new cells in the hippocampus and might significantly improve our ability to lay down new memories for as long as that level of neurogenesis (and exercise!) lasts. I don't know about you, but my memory-encoding abilities could always use some help, and understanding how this works and how to maximize this effect is one of the major goals of my lab.

Combining exercise and academics was an innovative element of my class, but turning the class into a real research study was even more exciting. This way, not only would the students become the research subjects but I could also include them in the act of analyzing the data. As part of the class, the students would have a chance to study the data from their own class (names removed, of course) and the control class to determine if exercise had really improved their memory function. While these students had all done lab courses, this kind of data analysis is generally done in a working research lab. We had brought the research lab to the classroom! What better way to apply the knowledge that they would be obtaining during the semester through my lectures and our discussions than to analyze the findings from a real experiment?

TRANSFORMATIONS CONTINUE

The intentional exercise I had been practicing not only inspired me in my teaching and research but was starting to shift the way I approached other parts of my life as well. After a little break following

Cabin Boy and Car Boy, I was ready for the next step. I had two main questions for myself. How do I build a richer social life and how could I make myself feel ready and open for a lasting relationship?

After striking out with my first matchmaker, I decided to try it one more time with a different matchmaker—after all, Car Boy and Cabin Boy were both reasonable on paper, maybe I just needed someone with a new and different male dating roster. I met this next matchmaker in the lobby of a very cool hotel, and she seemed to be connected to all the right people and said all the right things. I signed up, and she connected me with a businessman and a doctor, both fairly nice and both decidedly not for me.

Then I seemed to hit it big. Through the matchmaker, I met a very sweet and intelligent lawyer. He lived in the city, and his name was Art. This one stuck. Neither of us had dated very much in the past few years, and we were both eager to find a steady relationship. We made a great couple—for a while—enjoying dinner together, weekends at his place in New Jersey, and occasional outings to the theater. Then Art and I got to the stage of our relationship where our various differences became more and more pronounced. And I thought I had a possible solution.

I would hire a personal coach. When I wanted some help in the dating department, a professional matchmaker seemed to do the trick. Now that I needed help improving my relationship, maybe a life coach would help. My gym sometimes offered free life-coaching sessions, so I signed up for a thirty-minute trial with coach Marnie.

Marnie was amazingly perceptive and immediately started helping me understand how I was in my relationship with Art *and* how I was in my relationships in the rest of my life. It turns out, I had a lot of things to clean up in my relationships overall. I had been focusing so much of my attention on my work, I hadn't paid enough attention to the maintenance of my personal relationships. I learned quickly that it was not so much that I didn't know how to have strong personal relationships, I just needed to focus more of my attention on them and they would start to thrive in the same way my career was thriving. I also needed more than a little guidance to get this process started, which is what I got in spades from my coach.

What did this relationship cleanup process look like? My favorite example involves a surprising target of my mission: a doorman in the co-op building where I live in New York. One of my negative personality traits that I identified with my coach is that I am very easily insulted. Even worse, once insulted, I don't try to repair the hurt; I just stew. On top of that, I can hold a grudge for a very long time. You can imagine that a long-time grudge against my doorman, someone I rely on and see several times a week, not in my house but essentially outside my front door, is a prime candidate for cleanup!

The perceived insult happened very soon after I moved into my building. I tried to schedule a furniture delivery with the doorman on duty. But unlike my other friendly and helpful doormen, this one seemed quite curt in his interaction with me. He told me that he was not sure if he could schedule the delivery and that I had to check with the super to be sure. He seemed a little annoyed at my question and was not particularly helpful.

I then started to notice that he never seemed very friendly when I walked through the door. Even more annoying was that I saw him interact with other residents in what seemed like a much friendlier way than he ever acted with me. He never engaged me or went out of his way to help me. In the end, I started to dread seeing him at the door and interacted as little as possible with him because it was so clear that he didn't like me.

That next Christmas, I prepared my annual tips for all of the building staff. I just could not bring myself to give this doorman a tip when I dreaded seeing him at the front desk, so I decided he would not be getting a tip that year. I knew that was a drastic (some might say foolish, childish, immature—fill in the blank) move, but I did it anyway, and then immediately regretted it for the rest of the year.

When I mentioned this to my coach, Marnie, as one of the relationships I had to repair, she asked me if I had ever spoken to him about why he was rude that first day. I said no. Then she asked if perhaps I had been a little cold to him, which might have influenced the way he interacted (or did not interact) with me. I admitted that my behavior could indeed be viewed as cold. My coach made me realize that this whole time, I had been making up a story in my own head that this doorman did not like me, even though I had been the one giving him the cold shoulder. She reminded me that this was a key relationship to clean up for me because this guy was a critical part of my extended home. He, like the other doormen in my building, knew all about me: what take-out I ate, which dry-cleaning shop I went to, and which friends visited me and when. Doormen often know about your romantic relationships before your BFF because they are the first ones to see who arrives early and stays late and who doesn't go home at all. She said that the only way to turn this relationship around was to confess my theory about him to him, and see what he said. And she helped me work out what I was going to say. Yikes!

I remember sitting in my bedroom on the morning I knew he was working, going over my speech in my mind and really not wanting to go downstairs. I was experiencing that sick feeling that comes before you take your most important final or right before you go on stage for the first time. I somehow managed to stand up and make my way downstairs. When the elevator doors opened, I marched right up to him and said in a slightly trembling voice, "Hi. I wanted to ask you something. I first wanted tell you that I think the entire staff at this building is great and I appreciate all the great service. But, as you know, Christmas is coming up again, and I wanted to tell you that I felt so bad that I did not give you a tip last year at Christmas. I didn't give you a tip because I have this idea that you don't really like me, and I wanted to talk to you about this."

Yikes—I had actually said it.

He looked totally stunned.

It took him just a second to recover before he assured me he harbored no ill-will toward me. Instead he said that his style was to stay professional and stay out of people's business and suggested that that behavior might have been mistaken for him not liking me. He pointed out that he was very different from some of the other doormen who liked to talk to all the residents and ask questions about their lives.

My key realization was that he seemed genuinely surprised that I thought he did not like me.

I thanked him for his honesty and told him I must have just completely misinterpreted him. I reiterated that I thought he and the entire door staff did such a great job and I looked forward to rewarding everyone at Christmas.

This was maybe not the most elegant conversation I ever had, but I did what I had gone there to do. I thanked him again and awkwardly ran out the door. I nearly cried with relief as I went flying down the street to the subway, trying to get as far away from that stressful, yet successful conversation as I could.

But was it worth it?

Yes.

This embarrassing, awkward, difficult, and stress-inducing conversation completely shifted our relationship. Both of us now had permission and motivation to focus on positive interactions, and we were both about 300 percent more friendly every single time we saw each other after that, including that very night when I came back home. It was like my own little Christmas miracle.

This was just one example of how I was shifting the relationships in my life. More important, I was becoming more and more sensitive to the health of all the personal relationships in my life, and I was on a mission with the help of my life coach to fix them.

How did this relationship fixing work out with Art? I realized I had to ask myself why I wanted to fix this relationship. Was it because I loved him and wanted to spend the rest of my life with this man, or was it that I liked the idea of being with someone even if he was not particularly compatible with me. Art was very kind-hearted and very smart, both of which I appreciated. But in the end, we didn't share a great deal of lifestyle traits. Probably the most telling difference was that he had very few friends in New York, and he was not social at all. He enjoyed hanging out with me but was not interested in meeting or knowing my growing group of friends. In the end I wasn't sure he would like or get along with them very well anyway. I realized that while just a short time before, I had been just as isolated socially as he was, I had clearly and deliberately changed that in my life, and staying with him would be like moving back to those days of social isolation that I had worked so hard to change. He also had no interest in good food, which was challenging for me in the very beginning and another clear sign that we were just not meant to be.

It sounds like an easy decision to make, but it wasn't. I feared that if I ended it with Art, I would never find anyone else for the rest of my life. But the decision was basically made for me. We were just not getting along, and there was no choice but to break it off with him.

While the development of my new exercise class was an exhilarating part of the transformation I was going through, my breakup with Art was an equally important, and very painful, part of the same process. At the heart of this was me becoming more self-aware. That may sound obvious. But keep in mind how focused on science and my work I had been for the previous twenty years. Did I love my career? Did I love the work I did in my lab? The research?

The reading? Yes. But I was realizing that if I wanted a more full, balanced life, I had to literally rebalance my own brain–body connection, with more focus on the body part of the equation. Then I began to seek this same kind of balance in my relationships; I wanted more from them. This became a process of becoming more aware of what I really wanted in life. I had spent years focused on the goal of getting tenure and working hard, loving the science that I did but putting anything that got in the way of my job on the back burner. I focused so much on my work goal that I didn't pay nearly enough attention to the present moment. I still had a way to go, but this whole process was one of starting to pay more attention to my emotions and desires, my likes and dislikes, and being more open to what was going on at that moment instead of focusing so much on trying to control the future.

And that's part of what my regular exercise provided: a constant focus on the present moment. You really can't be thinking about the future when you're in the middle of a serious workout session, especially an intentional workout like I was giving myself. For those sixty minutes of intenSati class, I felt completely connected to myself: my thoughts, emotions, and physical movement. I was all one—body and brain. I was supremely aware of how I felt, how my feelings changed during class, and all my emotions that came during and after the workout, from annoyance to joy to peace to relief to exuberance. I experienced a moving meditation in which I was able to be in the present moment, allowing myself to experience what my brain and body were telling me.

This sensory, intensely emotional, physically charged experience became the catalyst to figure out what I *really* wanted to do with my life, with no expectations and no preconceived notions. What can bring me joy? What do I no longer need if it doesn't bring me satisfaction or joy?

TAKE-AWAYS: EXERCISE AND NEUROGENESIS

- Exercise is responsible for the majority of the positive brain changes seen with environmental enrichment, including increases in the size of the cortex, enhanced levels of growth factors like BDNF and neurotransmitters like acetylcholine, and enhanced growth of blood vessels in the brain (angiogenesis).
- Both exercise and enriched environments enhance neurogenesis, or the birth of new brain cells, in the hippocampus.
- The increased levels of BDNF stimulated by exercise help the growth and development of these new brain cells.
- Exercise also enhances the volume and size of the hippocampus, increases the number of dendritic spines on hippocampal neurons, and enhances the physiological properties of hippocampal neurons as measured by LTP.
- In humans, most studies of the effects of exercise have been done in the elderly.
- In studies in the elderly, higher levels of exercise are correlated with lower incidence of dementia later in life. This is an example of an observational study.
- In elderly people, randomized controlled studies (interventional studies) have shown that increased exercise can improve attention and increase the size of the hippocampus.

BRAIN HACKS: HOW DO I INCREASE MY EXERCISE? PART I

If you don't have hours in the day to exercise, here are some ideas that I use in my own life to get that regular exercise practice going in just four minutes:

- Walk up the stairs to your favorite upbeat song (like "Happy" by Pharrell Williams) until the song is over, then take the elevator the rest of the way up.

- Challenge a friend or coworker to do a combination of desk push-ups and squats at work for four minutes a day (or multiple times a day for a whole week). Then challenge someone else.
- While you are brushing your teeth for four minutes, do a rotation of deep knee squats and side bends (face toward the mirror and slowly lean your body to the right and then to the left, stretching out the ribs on the side opposite to the direction that you are leaning). To make the side bends more difficult, wrap a big towel on your head as if you were drying your hair; it makes your side abs work even harder.
- Keep it fun by playing a four-minute game of tag with your kids, or borrow someone else's kids to play it with.
- Set a timer for four minutes and clean up as much of your home or office as you can as fast as you can. Try cleaning the bathtub or doing some speed vacuuming or mopping; that can really work up a sweat and it will last only four minutes!
- Be a kid again by using a Hula-Hoop for four minutes. It's an amazing aerobic workout for your abs and core.

6

SPANDEX IN THE CLASSROOM

Exercise Can Make You Smarter

A t 9:25 A.M. on September 7, 2009, I was clad head to toe in my best spandex, standing in front of a room of undergraduate students at NYU, ready to lead my very first "Can Exercise Change Your Brain?" class. As we stood about in the same space where I had lectured countless times before, the students were decked out in a varied assortment of what could pass as workout clothes—from athletic to sassy to Goth to rumpled, messy, and pajamalike. Clothing was not the only thing different that morning: I felt nervous. And I had not been nervous before teaching a class in at least ten years. I had a lot on the line.

I had been training to be a kick-ass intenSati instructor for the prior six months, and had been planning this class for over a year. This day was the culmination of my vision to bring together two vastly different worlds: exercise and neuroscience. And I was going to do it in a way that had never been done before. Was I just wasting my time? I wondered if my colleagues thought I was more than a little off my rocker to put so much effort into this crazy exercise class when I didn't even study exercise in my lab (or at least I hadn't up until then). At that point even I considered it my pet science hobby. It was definitely risky to spend so much time and effort on this one class. I had some doubts even as I was developing the class syllabus. In general my department head had been open and tolerant, allowing my colleagues and me the room to develop a wide range of different courses, but no one had designed a course like the one I was about to launch.

Apparently, I wasn't the only one who was nervous. While all the students knew that they were going to exercise in this class, I could tell they did not quite know how to react when confronted with their spandex-clad professor. When I looked out at the room, I saw a mixture of expressions: fear, amusement, querulousness, studied boredom, and hints of nervousness. Well, I would never know what was going to happen unless I actually started, so I calmly stepped out in front of the room and said, "Welcome to the very first 'Can Exercise Change Your Brain?' class! I hope you guys are up for something a little bit different because NYU has never offered a class quite like this before. This course was inspired when I decided I wanted to get in shape, and I started going to the gym. I became a regular gym goer and noticed how much exercising really helped my attention, my energy, and my concentration at work. With that, the idea for this class was born. I was fascinated with the neuroscience underlying this change, so I designed this course in which we will be combining physical exercise class with lectures on the effects of exercise on the brain. And here we are."

I told them that we were not going to be just passively learning about the neuroscience research that has examined the effects of exercise on brain function; instead we were going to be actively participating in the research process. In fact, they were going to be tested on their performance at the beginning and then again at the end of the class, up against a control group, to see if exercise could *really* change their brains. We would also be considering lots of different questions, such as how exercise changes your brain, how much exercise is needed to make a change, how long does the duration of exercise need to be (days vs. weeks vs. months), and what kind of exercise might elicit the best response.

Then I asked, "Okay, are you ready to work out?"

A quiet murmur of consent followed. Not the level of enthusiasm that I was looking for that morning.

So I repeated myself, dramatically cupping my hand to my ear as I asked, *"Are you ready to work out?!"*

They came back with a strong yes. And with that, we started exercising.

SWEATING IN THE CLASSROOM

I started the class by explaining how the exercise part was going to work and that we would be pairing movements from kickboxing and dance and yoga and martial arts with positive spoken affirmations like "I am strong now!" and "I believe I will succeed!" There were a few incredulous looks and giggles that spread over the crowd with this explanation, but once the giggles died down, I turned on the music. I needed to start.

With music blaring from the surround-sound speakers in our classroom, I introduced the first move, one called Commitment. I showed them how to stretch their right and left arms up in the air (hands wide open and fingers spread) in an alternating fashion to the beat of the music.

Once they got that, I added the affirmation: "Yes! Yes! Yes! Yes!"

That was easy. The next move was called Strong, and was accompanied by alternating right and left punches with legs apart and bent at the knee in a light squat.

The students picked up the affirmation "I am strong now!" quickly, and the rest of that day the room rang out with exclamations of:

"I want it, I want it, I really really want it!"

"I believe I will succeed!"

"I am ready to be inspired!"

"I am willing to be inspired!"

"I am able to be inspired!"

"I am inspired right now!"

As we moved, bursts of laughter peppered the classroom, but it was joyful, playful laughter as if the students could not believe they were actually jumping around, sweating, and yelling out affirmations in the same classroom where they usually spent all their time sitting and listening to neuroscience lectures. I could hardly believe it myself.

All my butterflies disappeared as I got into the flow of teaching this exercise class.

I asked them questions like "What are you committed to?" and "What are you saying yes to in your life?" when we were doing the move Commitment.

I asked, "When you do feel strong in your life?" when we were doing the move Strong. Sometimes I stayed on a particular move for longer and did different variations to get the students moving around the room or have them change places in the room. The key was to be very systematic about how I taught the moves and affirmations but to also keep it interesting by adding variation and by asking them to think about the affirmation. As we learned and performed the series together, I traveled the classroom, mirroring the movements right in front of the students or calling individual students out by asking,

"Becky, are *you* ready to be inspired?"

"Ed, are you strong now?"

This was a great way to engage students in the workout, and it forced me to quickly learn their names. As an academic lecturer I always thought I did a pretty good job at engaging my students and trying to ask relevant questions and encouraging discussion in class. I could already tell on that first day that interacting with the students in this way was changing how I taught. I was discovering a whole new way to relate to them as students and I learned it . . . at the gym!

IT'S ALL ABOUT THE STUDENTS

In memory research, there is something called the primacy effect. It refers to how our brains remember the first items in a series very clearly and strongly. We all have experienced this phenomenon: Think about the immediacy of the memory of your first date, your first kiss, and your first day at a new job.

When I began teaching my new class I realized that I was smack dab in the middle of lots of firsts. This was my first time ever teaching an exercise class. My first time deeply exploring a topic outside of memory and hippocampal function in front of a class full of students. My first time venturing outside of my comfortable "I am an expert professor in charge" box. Yes, this was a brand-new day, and I was immersed in a completely novel experience.

You had to have a bit of an adventurous spirit to sign up for a new class, and my students certainly fit that mold. Jamie was a standout from the beginning, very engaged and a real leader in the classroom. She was there because she had a personal interest. Her autistic sister had been greatly helped by exercise; in fact Jamie told me her mom had always had her sister on a serious exercise regime as part of her therapy. Maybe because of this emphasis on exercise in her childhood home, Jamie herself was not only very athletic but also racked up some of the highest weekly exercise hours of the entire class (I had all students in the class keep a weekly exercise log). She had also worked for wilderness camps for autistic kids and had seen for herself the positive effects of physical exercise on autistic symptoms.

Jamie wanted to learn more about the research that supported what she had seen and thus had a deeper purpose in taking the class. She was not there just to fulfill a requirement and get a good grade. Her focus helped set the tone for the entire term. By the way, Jamie went on to graduate school to study, what else: autism.

Emily was the giggler of the group, leading the infectious laughter during the workouts throughout the semester. I still remember the first day of class when she was laughing so hard her glasses steamed up and she could not see. I found out later in the semester that Emily was also working in the lab of a colleague at the medical center, examining the effects of obesity on cognitive function in adolescent and preadolescent children. That work, headed up by Professor Antonio Convit, provided some of the first evidence that obese adolescent children with type 2 diabetes perform worse on a range of different cognitive tasks than do normal children, suggesting that obesity with diabetes is not only terrible for your health (adolescents with type 2 diabetes can expect to live twenty years less than the rest of the population), but bad for your brain as well.

Given her bubbly personality, it was not surprising that Emily graduated that year and started working for Teach for America, where I know she is bringing that same lightness and levity, along with the neuroscience of exercise, into her own classroom.

The academic part of my new class included a lecture (typically thirty to forty-five minutes long), during which I went over material from the readings that I had assigned. We started with the early, more historical precursor studies before launching into neurogenesis and the effects of exercise across different species. The last part of class was my favorite—the discussion session. Here I challenged students to not just read the articles I assigned but also to propose a new experiment based on the readings and current research. This shifted the class from a read, recite, remember strategy to one in which I asked them to think like scientists and ask interesting scientific questions. For example, the journal articles that I had assigned might describe a series of experiments in which scientists gave rats access to a running wheel to increase their exercise and then measured neurogenesis (the birth of new brain cells) in the rats' hippocampi. The articles described the relationship between neurogenesis and the rats' improved performance on various memory-demanding tasks. That week in class I asked my students to come up with an original experiment related to the studies they had read. They might propose to look at the effects of exercise on other brain areas like the prefrontal cortex and describe the tasks they would use and an experiment to start to understand what brain changes might underlie any improvement in prefrontal function with exercise. Or they might propose to examine a particular molecular pathway that might be involved in the change in brain function that the papers described. The great part of investigating the effects of exercise on brain function is that it's a relative new topic in the field of neuroscience (a baby of a topic compared to memory, for example), and while there is a solid base of studies to learn from, there are also some key basic questions that remain unanswered, as I'll discuss later in this chapter. My goal was to get the students first to identify the key questions and then to start to imagine and develop new experiments to address the questions.

This kind of assignment asked for a different kind of thinking and analysis than they were used to from other classes. I wanted them to use what they learned from the research findings to generate the next interesting question. While it was designed to be fun, it ended up causing anxiety in at least some of the type A students. I remember Becky had just presented her new experimental hypothesis in class one day when she blurted out, "I *know* this is bad, so don't even tell me it's good." I laughed and told her that learning how to ask good experimental questions was a process and requires courage—courage to fail especially. I also assured her that knowing when a question misses the mark is a valuable first step toward figuring out what does make a good research question.

Slow but steady improvement paid off like gangbusters for Becky, who by the end of the semester turned in a beautifully designed experiment examining the specific molecules involved in how exercise enhances neurogenesis. She told me afterward that she could not believe what she had accomplished and that the combination of copious encouragement and gentle critiquing I had done had helped her immensely. I loved seeing this success, and it proves a major belief of mine: Anyone can learn to think like a scientist!

WHAT REALLY HAPPENED WHEN I BROUGHT EXERCISE INTO THE CLASSROOM?

My original idea for bringing exercise into my classroom was that I wanted students to feel the high they got from a great workout as they were learning about all the effects of exercise on the brain. While I also hypothesized that exercise would give them a boost of energy that would also affect the academic part of the class, I never could have predicted all the changes that actually occurred—not only for my students but for me as well.

I used intenSati as the workout for this course for a few reasons. First, as I've described, I had found intenSati to be an incredibly motivating and uplifting workout, and I wanted an engaging form of exercise to keep the students motivated and involved in the exercise component of the class throughout the semester. But there were also reasons *not* to choose this workout. intenSati involves not only an aerobic component but a strong motivational component as well. I strongly suspected that the motivational/affirmation part of the workout added mood-boosting potential beyond the exercise itself. I weighed the pros and cons and decided for this class, keeping the students motivated and engaged in the exercise was most important so I chose to stick with intenSati, but I knew that for anything we found, we would have to go back to determine if exercise alone, affirmations alone, or the particular combination of intentional exercise was causing the effects that we saw. Indeed, this form of exercise did bring a whole new level of energy and excitement into the classroom and everyone, including me, felt it. This positive energy from the exercise portion of the class easily seeped over into the academic lecture and discussion parts. The most obvious change was that the students were more energetic and completely and totally awake after the workout when we got to the academic part of the class. We had just spent the last hour sweating, and working, and high-fiving each other during the exercise, which put the students in an excellent state of mind to learn: They were relaxed but aroused, focused and attuned, and the topic seemed highly relevant and interesting to them.

In addition, I ended each intenSati session with a three-minute meditation (see Chapter 10). This allowed the students to quiet their minds before we started the academic part of the class.

This intentional aerobic exercise together with the short meditation seemed to provide the students with both energy and focus for our academic learning session that followed. In fact, one of the students said this class was nothing like her other morning classes, where she was clinging to her Starbucks cup. Instead she said she was able to remember everything from this class and didn't even need to take notes. Another student said she felt like she was able to pay better attention in this class relative to her other classes.

Part of this energy boost I believe came from the rather unusual situation of having students work out with their professor (in fact, led by their professor) before class. But in addition to this novelty, I explicitly challenged the students to be as interactive in the lecture part of the class as they were in the call-and-response part of the exercise class. All of these factors worked to give the students a higher level of energy than I'd ever seen before in my classroom. I believe the positive affirmations played a role in this positive shift in energy. As I described in Chapter 4, not only have positive self-affirmations been shown to buffer us from stress and improve our mood in certain situations but aerobic exercise increases a wide range of different brain hormones and neurotransmitters like dopamine, serotonin, endorphins, testosterone, and BDNF, all of which have been shown to have a positive effect on mood. Consistent with these findings, the mood in that classroom was somewhere up in the rafters.

But of course the students were not the only ones benefiting from the exercise. I was too. The first thing I noticed in me was the boost of energy that came from leading the exercise part of the class. I am typically tuckered out after teaching a traditional class, and I was worried I wouldn't be able to make it through the lecture right after having taught an hour of exercise. I found instead that I was more energized after the entire combination of exercise plus lecture/discussion than I was after my typical lecture classes. I was getting a boost of mood from the exercise and affirmations and from the increase in testosterone from the powerful poses I was leading the students in.

The most striking change, however, was how differently I engaged with the students in that class. The exercise class with all its shouted affirmations and playful interaction, which was not part of my regular academic classes, spilled over to the lecture/discussion parts of the session, so I was interacting with students during the entire class in a much more relaxed way. I also shared more of myself with the students in this class, finding myself openly and easily telling them stories about how the affirmations were working (or not) in my life. These were not science stories, but personal examples of how I used persistence (one of the affirmations) in my life or when I had been strong. This was part of the training I received to be able to teach the exercise class, but I had no idea how much that personal touch would change the interactions with the students in the classroom. It reminded me of the personal stories about her life or her family that Marian Diamond would share with us during class. I remember her stories, but it took me years and years to understand how important they were to establishing the kind of wonderful rapport she had with her students.

MY TEACHING SECRET WEAPON

One of the best and most powerful teaching tools to make something memorable to students is novelty or the element of surprise. Something that is novel or surprising focuses attention, engages emotional systems, and is therefore highly memorable or "sticky," as neuroscientists like to say.

Let me explain.

I was once teaching a 120-student core curriculum course called "Brain and Behavior" for nonscience majors, and we were coming up to my favorite part of the syllabus, the classes on memory. I wanted to introduce this section in a very memorable way, and I started thinking about ways that I could do that. At our next teaching meeting I introduced this idea to my teaching assistants (TAs) and then asked them to brainstorm. Erik had a fantastic idea. He said he knew a graduate student in our program who moonlighted as a popular underground burlesque performer with the stage name Dr. Flux (remember, I work in New York City!) and one quick e-mail confirmed that he would be happy to participate in our little scheme.

On the day of the introduction to memory lecture, I started as I usually do, describing the basic concept of memory. Suddenly Dr. Flux burst through the door in a nerdy-looking suit covered by a lab coat, surprising everyone, including me. I was surprised because I had not seen him in full costume before and half of his shaved head was covered with an inch-thick layer of what looked like gold sparkle dust. On cue one of the TAs turned on some music, and

Dr. Flux started strutting in front of the classroom, doing a risqué dance number. At this point the students were turning to one another and to me, wondering what was going on. Then Dr. Flux took off his lab coat and ripped off most of the rest of his clothes (held together by Velcro) under which he revealed a rather small pair of bright gold booty shorts (he had failed to mention just how tiny the shorts were when he had first described his act to me).

Dr. Flux then began undulating to the music right next to me at the front of the classroom, at one point even doing a back bend. Finally, he pulled on a bright orange hazard suit and dramatically left the room.

You should have just *seen* the look on the students' faces. After Dr. Flux walked out the door, without missing a beat, I turned to the class and asked, "So what makes things memorable?"

One young man way in the back of that big classroom, who had never raised his hand before, put his hand up. I called on him and he said with certainty, "Gold booty shorts!"

I said, "*Yes!*"

I explained that the demonstration used to start off our section on memory was in itself memorable because it was novel (gold booty shorts), surprising (gold booty shorts), emotional (stripping to reveal the gold booty shorts in the context of a classroom), and attention grabbing (gold booty shorts). I guarantee that those students will remember that introduction to memory for a very long time.

CAN EXERCISE CHANGE YOUR BRAIN? THE RESULTS SHOW

The great thing about the "Can Exercise Change Your Brain?" course is that I had more than just anecdotes to tell about how this class was so different from any other class I taught; I had actual experimental evidence. As you will remember, in this class, we asked the question, Could once-a-week aerobic exercise (plus affirmations) for a semester (fifteen weeks) improve memory encoding relative to students in a different elective neuroscience class that didn't include exercise? At the end of the study, we had relatively small numbers of students in the exercise class and the control class. The deck was stacked against us because of both the small number of subjects and the low number of exercise sessions they completed during the study.

We examined the results of the memory-encoding task (which is supposed to depend on the part of the hippocampus where new cells are born) to see if there was any difference in the performance of the two classes. We were all thrilled to find a significant improvement in the exercise group relative to the control group in one measure. We found that the exercise group responded to the correct stimulus in the memory-encoding task significantly faster than the control group. Processing speed is one aspect of cognition that has been reported to improve with exercise, and we were able to see this in this study.

Now, previous studies had shown that rats performed significantly better on a task after exercise, just like our memory-encoding task. We didn't see an overall improvement in performance (that is, more correct choices), but one big difference was that the experimental rats got much more exercise than the control group rats—sometimes running ten kilometers a day (rats love to run). We didn't see an overall improvement in memory in my students, but we seemed to see just the first hint of an improvement on the task in the form of reaction time. This was important because it told me that with just once a week of increased exercise, you can start to see a significant effect on some measure of the memory-encoding task in healthy young adults.

This finding lit a fire under me like no other result I had gotten before because it suggested something very exciting. If we could start seeing reliable effects in healthy young adults with just once-a-week exercise, what would we see if the subjects increased their exercise to three or four times a week? Now *that* was exciting and is among the questions we are pursuing now.

Now, the other question from this first study was whether the effects we saw were due to the increase in aerobic exercise alone, the positive affirmations (the intentional component) alone, or the combination of the two. As I mentioned, there was a huge motivational shift in terms of interaction in the course, and many of the students commented on how the affirmations that we said in class stayed in their head all week. We focused on the mood component of inten-Sati in another study I did soon after the exercise study in students, this time focused on a patient population group: individuals who had suffered traumatic brain injury (TBI).

EXERCISE EFFECTS ON MOOD IN TRAUMATIC BRAIN INJURY

A collaboration that developed after a talk I gave at NYU's medical school campus about my exercise class was with Dr. Teresa Ashman, from the Rusk Institute of Rehabilitation Medicine at New York University Langone Medical Center. Ashman specializes in the rehabilitation of patients with TBI. This seemed like an ideal population to try our exercise intervention on for a variety of reasons. First, these patients suffer from a range of cognitive deficits, including difficulty with attention, motivation, and memory. Depression and fatigue are also common in TBI patients. We reasoned that long-term exercise might have the potential to improve the cognitive problems seen in these patients as well as improve or alleviate the depressive symptoms. So we set about to design a simple experiment in which we tested subjects before and after an eight-week exercise intervention that consisted of twice-a-week group exercise. Control subjects did no exercise for the same time period.

I went to the first session of the experiment after the participants had been chosen for the study, ready to give them all a big welcome and pep talk, to tell them why we were doing this work, and to encourage them to the best of my ability to come to as many of the exercise sessions as they could over the next eight weeks. The age of the participants ranged from the twenties to the sixties (you can suffer from TBI at any age). Luckily, I met a positive and expectant attitude in the room that day, and I was confident that we had a great and engaged group of subjects.

I then asked the exercise instructor, Amanda Berlin, to lead the whole group in a short and easy three-minute exercise demonstration. The participants looked a little unsure about standing up and exercising at the introductory session, but all joined in eventually and seemed to enjoy it. Well, almost everyone. When the demo was over, I turned around and saw a young woman walking toward me with a pinched red face who looked like she was about to cry. Her name was Angelina. I immediately asked her what was wrong. "That was torture for me!" she spat out. "The music was way too loud and too fast and the bright lights in the room really hurt my eyes!"

I could tell that Angelina really wanted to be a part of this study, but she was so frustrated by the demonstration that I worried she wouldn't return. Ashman and I assured her that we would make sure that the music was slower and the lights lower next time, and that she could do the workout sitting down if that was easier. We then introduced her to the instructor, and assured her that Amanda would take great care of her. Eventually, Angelina calmed down.

During the eight weeks that the test participants were meeting for exercise class, I got regular updates on attendance and made sure that the instructor had all the support she needed to run the classes efficiently and easily. At the end of eight weeks, I was touched to get an invitation from the participants, who invited me to come to their last class so I could see their progress.

I arrived in the NYU medical school classroom and immediately noticed that the room was buzzing with energy. Everyone was excited to be there, and the participants were all saying how they could not believe how quickly the eight weeks had passed. When everyone had arrived, Amanda started class, and I could not help but notice that her music was as fast as the music in any local gym in New York City. Not only was the music fast but everyone was keeping up. I saw the group jumping and punching in perfect time to the music—they clearly knew what they were doing.

But it was what I saw in the front row of the class that made my day, week, month, and year. It was Angelina. I had hardly recognized her, but there she was, with a huge smile on her face doing the workout like a pro. I looked twice to be sure that it was her, and sure enough, it was! I spent the whole rest of the hour marveling at what I was seeing.

After the workout and meditation were over, we all sat in a circle. I asked, "Do you all know how amazing you are?" The whole group was beaming.

I continued. "Do you know what a difference I see from where you all started eight weeks ago? What happened?"

They all then started pointing at one another and talking all at once. One beautiful young woman said that she learned how to smile again in this class, and she had also invited her therapist to come to this last class so she could see. Another woman said that she was just so inspired by seeing the improvement in everyone else each week. They all credited Amanda for being a wonderful leader. Angelina described her amazing transformation, which was actually a gradual but continuous process. She said the first week of class she could not even feel her feet under her body during the exercise, but by the second week she could. She could not do both the arm movements and the leg movements at the same time so she worked on one at a time. But soon enough, everything just came together, and after regular attendance, she could suddenly do all the movements—all at once. I can't describe the look of joy and accomplishment that she had on her face as she spoke.

A real transformation had taken place in that exercise group, and it was one of the most beautiful things that I had ever seen. And the study was not even over yet. The participants had all taken cognitive and mood tests at the start of the exercise regime. Luckily we had really focused on this aspect of the change and had a wide range of different mood and quality-of-life assessment forms we had given them to fill out. Now, they would retake all the same tests to determine if they had changed relative to the nonexercising control group. What we found reflected exactly what I saw in that final exercise class. There was significant improvement in mood and quality-of-life measures for those who had done the exercise class compared to the TBI patients who had not. As measured by the various surveys they took, the exercising TBI group had decreased scores on a depression and fatigue index and increased scores for positive affect and quality of life. I got a glimpse of all of those changes when I saw the group in exercise class. It turned out that none of our measures of memory or attention had changed, but this might have been because, although the group eventually worked themselves up to quite high levels of aerobic activity, it took several weeks to get there. In other words, while we saw clear improvements in mood, we might not have had an intense enough level of aerobic activity over the eight weeks to see improvements in cognitive functions.

The paper Ashman, our research team, and I wrote after we completed the TBI study turned out to be my first published report about the effect of exercise in a patient population group. What a thrill! Our findings suggested that eight weeks of twice-a-week exercise could significantly improve mood, positive affect, and quality of life measures and decrease fatigue in TBI patients. While our study size was small, this was an exciting result. But then, we had to ask ourselves, What was really underlying this effect? Was it the exercise? Was it the fun and interactive group environment? This study alone could not answer those questions, but future studies can be done to tease those important factors out. The important point is that our results suggested that intentional exercise can improve a range of mood and fatigue measures in patients with TBI. I was more than happy to take that as a start.

TELLIN' IT LIKE IT IS

I typically get one of two reactions when I tell people that I'm a neuroscientist who studies the effects of exercise on brain function. The first is "That is *so cool*—I want to know all the results of your studies!" The second is "Of course we know that exercise improves brain function! Isn't that old news?"

I think these two responses reflect the effects of the popular press on this exciting field right now. On the one hand, there are popular articles being published almost every day about the positive effects of exercise on brain function. These articles are wonderful in the sense that they keep the general public up to date on the latest findings, but they have a tendency to make strong conclusions based on the publication of a single study, giving the false impression that we know much more than we do. So I can understand the people who have the impression that most everything is already known.

The reality is quite a different story. It's true that there are more and more neuroscientists focused on examining the effects of exercise on brain function in both animals and humans, but there are still many big and exciting questions left to address. For example, the studies thus far in animals have focused strongly on the effects of exercise on the hippocampus and on identifying the neurotransmitters and growth factors that change with exercise. There is a rich literature there, but an exciting direction will be examining the brain effects of exercise outside of the hippocampus. For example, the most common finding in humans is the effects of exercise on prefrontal function. Very little is known about the effects of exercise

in the prefrontal cortex in rodents. Similarly, there is exciting literature on the positive effects of exercise in the easing of symptoms in
Parkinson's disease. These findings suggest a strong effect of exercise
on the striatum, the primary area of the brain damaged in Parkinson's. Very few studies have examined how exercise affects the striatum in normal animals. But perhaps one of the biggest unanswered
questions that can be addressed by studies in animals is understanding the precise pathways, molecules, and mechanisms for how exercise is triggering any change in the brain. In other words, we know
that if you let a rat exercise, you can see changes in BDNF, endorphin, dopamine, acetylcholine levels, and neurogenesis, but we don't
know exactly how exercising is stimulating all these changes. It could
be myriad different factors that do the trick, and different factors
could be responsible for the wide range of observed brain changes.
It could be the increased heart or respiration rate or the change in
blood flow, muscle activity, and/or body temperature that stimulates
the brain. It's a complex problem and many basic questions remain
unanswered. One recent study claims to have identified a factor secreted by the muscles that can pass into the brain and stimulate the
release of BDNF. This is an exciting new report and will need to be
replicated and confirmed by other studies.

In people, a mountain of questions remains; many of them have
been raised by animal studies and are waiting to be confirmed with
randomized controlled studies in humans. The preliminary studies
have whet our appetite and told us there is something there, but
I want to know if all the striking changes reported in the rodent
brain can be seen in the human brain. In a nutshell, it's the prescriptive piece that is missing from the humans studies and that forms
the core of the research program in my lab. Here are some of the
key unanswered questions about the effects of exercise in humans
that fascinate me:

- How much (or little) exercise do I need to see improvements in memory or attention? **Answer:** We know that short-term exercise for thirty to sixty minutes can improve attention, but we don't yet know how long the improvement lasts. After an increase in exercise over eight to twelve weeks, we see an improvement in attention and sometimes memory.
- How long do brain enhancements last in humans after exercise? **Answer:** We don't know in the case of acute exercise or long-term exercise.
- Which brain function improves with the least amount of exercise? **Answer:** We don't know.
- What kinds of exercise are most effective? **Answer:** There is evidence that aerobic exercise is more effective than stretching or resistance training, but we don't yet have a good idea of which kinds of aerobic exercise might be best or what level of cardiac output might be best to enhance cognition.
- Does yoga help my brain? **Answer:** There are a few studies on the effects of yoga on the brain mainly focusing on how the meditative aspects play into brain function. But there are too few studies of this kind to make firm conclusions.
- Can I take a pill and get the same effects as exercise? **Answer:** No. While many have attempted to create this magic bullet, there are no pills available that can reproduce the widespread effects of exercise on brain function.
- What time of day is best to exercise? **Answer:** The first answer is any time of the day that lets you exercise regularly! The scientific jury is still out on definitively determining the best time of day to exercise. My personal preference is to exercise first thing in the morning. This gets the positive hormones, neurotransmitters, growth factors, and endorphins flowing, preparing me for the workday. While this may be true, it

still has not been definitely proven which time of day is more beneficial for cognitive performance. It may also be that no matter what time the workout, the long-term changes in brain chemistry and function may emerge irrespective of the time of day of your regular workouts. In the absence of definitive evidence, I choose the theoretical best answer for preparing the brain for a day of work: an early morning aerobic workout.

All the exciting research findings in rodents showing striking brain changes conferred by exercise make exercise one of the most exciting potential therapies around. It's free and available to all. It has the potential to improve brain function in healthy brains, young brains, old brains, and diseased brains alike. It can help students young and old learn better in school. And it makes you happy as it improves various cognitive functions in your brain. This is why I'm excited to devote the next phase of my career to this area.

THE CLASS THAT REFLECTED MY LIFE

My "Can Exercise Change Your Brain?" class, a brand-new kind of hybrid course, represented the brand-new kind of challenges I was taking on in my own life. Not satisfied with continuing on my single-minded academic science track focused on the neurobiology of memory, now that I had tenure, I was starting to explore other areas of scientific interest to me irrespective of whether I had studied them before or not. I think I was just on a roll that started with the change in diet and exercise that I made in my life and how it had changed the way I felt in my body and the way I moved around in the world. It turns out that those changes also changed the way I saw myself. I started to see myself as strong, powerful (just like the affirmations that I was doling out in class), and able to make any change I could come up with in my life. And I was getting a little hint of that inspiration every single time I went to exercise class. In every class, in addition to the surge of all those good brain chemicals, I was reminded that I could push myself exactly as hard as I wanted to and feel the benefits in terms of my own strength and endurance. And this transferred to other exercise classes I was taking—from kickboxing to cardio sculpt classes to spin. And though these other classes didn't incorporate the same kind of explicit intention practice, my own awareness and attunement to my emotional experience during those workouts was just as elevated as during intenSati. With all of this mental and emotional energy in the form of heightened motivation, I felt I could move mountains.

This inspiration not only affected the courses I developed but started affecting other aspects of my life as well. For example, while I started out developing the "Can Exercise Change Your Brain?" class as a kind of science hobby, there was a clear moment while teaching that first semester when I realized the topic was more than just a hobby for me. It happened when Omar, one of the students in that first class, came to talk to me about doing some independent research on exercise in my lab. Omar was a varsity athlete. In fact he was the starting point guard on NYU's men's varsity basketball team and no stranger to long, hard workouts. He was quiet in class but had a deep interest in the brain effects of exercise because of the time he spent in the gym practicing. Soon after class started he came to my lab asking about whether he could do a research study under my guidance in the lab. While the study we were doing as part of the class focused on the effects of long-term exercise (increased exercise for three months), Omar was interested in asking if you could see evidence of significant improvements in cognitive functions after just one hour of aerobic exercise. It was a great question and I happily agreed to have him join the lab to do this project. But I actually had no other research in my lab at the time studying exercise. His request made me realize I did want my lab to start to study this question in a serious way. This was the moment when I left exercise behind as a science hobby and declared it a major theme of my research lab. I was not an expert in this area but was more than eager to do the work to become one.

This newfound spirit of exploration even affected my dating. I had had enough of the NYC matchmakers and decided to see what I would find online. They say that you attract the kind of person that you are yourself, and one day as I was browsing the profiles of an online dating service I had signed up for, I stumbled on one that immediately caught my attention. It had no picture, which usually meant an immediate pass for me, but the profile itself intrigued me. He was a professional musician, never married, living in New York, and had played with some of the best orchestras in the city. Hmmm. Ever since François, I've had a soft spot in my heart for musicians. Maybe this one was worth another look. For this particular online dating website you had to answer more questions to get to the next level, and when I did that, I did see a picture and I thought he looked nice, though somewhat intense. But he was a professional musician working in New York City, what did I expect?

THE MATCHMAKER FROM HELL

I don't have anything against matchmakers. The two I worked with provided me with some reasonable if not completely stellar dates and one serous relationship (Art). But my last encounter with a matchmaker put me off them for the foreseeable future. I went to an evening spa event with my dear friend Cheryl Conrad, a neuroscientist from Arizona, to hear what the guest speaker, a matchmaker, had to say. The matchmaker was very thin, and I couldn't help noticing that she didn't smile very much and didn't look all that happy to be there. She talked for fifteen or twenty minutes about her philosophy in matchmaking and the mechanics of how it all worked. It got interesting only after she finished her presentation and said she would take questions. An elegant-looking woman wearing high black boots and what looked like a Diane von Furstenberg wrap dress raised her hand, and said, "I have a friend who has a problem because she is always intimidating the men she dates. What do you suggest?"

Without skipping a beat the matchmaker said, "Well, the first thing I would do is stop wearing those hooker boots." Apparently, the matchmaker was referring to the high black boots that went a little above the knee that the woman was wearing. I would never have thought to compare them to the boots Julia Roberts wore in *Pretty Woman*, but that's me.

The room went deathly silent. I think everyone was holding their breath to see who would say what next. Fistfight? Screaming? Hair pulling? Anything could happen.

The woman calmly replied, "I said it was a friend."

After a split second, in which I do believe the matchmaker realized just how insulting she'd been, she said, "Well, intimidation is common in dating—especially in New York." I can't remember the rest of what she said because I was just so shocked at how rude she was and wanted to get away from there as fast as possible!

No, there were going to be no more matchmakers for me, and definitely not Miss Hooker Boots. Onward and upward!

After a week of answering questions about what we liked to do and where we liked to eat, we decided to meet for lunch near Carnegie Hall, where he often performed and practiced. We hit it off right away. While he was on the quiet side, he was very smart, interesting, and loved good food—a huge plus for me. This was starting to look promising.

Daniel and I went out for eight months, during which I got to hear a lot of fantastic music, eat at some great restaurants, and learn a lot about the classical music scene in New York City. Some of my warmest memories of our time together were the evenings I would get to see a beautiful opera performance (albeit alone because he was playing with the orchestra) and we would meet afterward for a romantic late-night dinner, and he would tell me all about the trials and tribulations of working with this difficult conductor or that one or about that one who completely lost the whole orchestra in the middle of the rehearsal. I loved it. But then again, on other nights, I would go home alone because he had to rehearse after the performance for the next day. At first his intense rehearsal schedule didn't interfere at all because I was busy too, but I came to realize just how consumed he was with his work.

In the end, Daniel wanted to get involved only so far. He was just not able to spend enough time with me because he was more married to his music than he could ever be to me, and we broke it off. It's always sad to break up, but something was different with this breakup. I realized I had enjoyed the ride. I was glad we went out and was glad to have known him and seen at least a little bit inside the world of the intense professional musician. I was getting both bolder and much more comfortable and confident in my own skin. This was clearly changing my romantic relationships, and I was able to step back just a little and appreciate it for what it was and move on. Don't get me wrong—it still really hurt when we broke up, and I had sadness and tears and anger. But that cleared up fairly quickly, and I was okay.

The other observation I made with more than a little bit of irony is that I was still a firm believer in the idea that you attract the kind of person you are. In Daniel, I attracted an intelligent, sensitive, interesting workaholic who was so obsessed with his work that he could not fit a close romantic relationship into his life. I had to look at myself and ask, Is that what I am? I had come a long way from the days when my social life was like a deserted ghost town from a Clint Eastwood movie. I really had. I had lots more friends and a much more active social life. But in the end, I still prioritized work over everything else (even if I made more time for those other things). And maybe I still judged myself by my success at work. I may have come a long way, but I still had a ways to go. And I would know I had gotten there when I started to attract men and other friends in my life who were not distracted workaholics but who were confident in their own skin and balanced work and other aspects of their life in the beautiful way I aspired to. Still a work in progress. But progress had been made.

TAKE-AWAYS: HOW EXERCISE MAKES YOU SMARTER

- Aerobic exercise can transform an academic classroom.
- One semester of just once-a-week intentional exercise can improve response time in healthy college students.
- Eight weeks of twice-a-week intenSati improved four measures of mood and well-being in patients with traumatic brain injury.

BRAIN HACKS: HOW DO I INCREASE MY EXERCISE? PART II

It's easier to get in your four-minute workouts if you involve your friends and family, so don't be shy about challenging them to join in the fun.

- Have a four-minute pillow fight with someone you love.
- Do jumping jacks through all the commercials of your favorite show each week and challenge your family to do the same.

- Challenge someone to an arm-wrestling match.
- Dance around your office, bedroom, living room, kitchen to your favorite song. (Try "Bang Bang" or "All About That Bass" for starters.) This is also a guaranteed mood booster and energy lifter—let those singers do the work to get you moving! If shadowboxing is more your thing, do that instead!
- At work, go to the bathroom on another floor and take the stairs.
- Bring a jump rope with you, and jump wherever and whenever you have time.
- Play and move with your dog or cat.

7

I STRESS, YOU STRESS, WE ALL STRESS!

Challenging the Neurobiology of Stress Response

D o you remember when you were a kid in school and you heard the teacher unexpectedly say the words *pop quiz*? Do you remember your heart starting to race and your palms getting sweaty as you waited for that dreaded sheet of paper to be passed down the row? That was your stress system at work. It turns out that little jolt of adrenaline you felt helped you remember all those state capitals better than if you hadn't gotten it. These brief bursts of stress are your body's nervous system becoming aroused.

Stress, in other words, is not all bad. Research has shown that moderate amounts of stress can be beneficial for our health, strengthening our immune system and cardiovascular system and speeding recovery from injury. Our stress systems also work as a vital warning system to help us get out of danger when we need to flee from a burning building or jump out of the way of a speeding car.

On the other hand too much stress, especially when it lasts for a long time with no end in sight, can be dangerous for our health. Chronic, long-term stress has been linked to heart disease, depression, cancer, and other life-threatening diseases. Like every-body else, I endure a wide range of different kinds of stressful situations each and every day. Everything from my daily commutes to the onslaught of a hundred new e-mails in my inbox to fighting the crowds at the supermarket is a source of stress. Stress is constant, inevitable, and seemingly unavoidable—or is it?

WHAT HAPPENS WHEN WE STRESS?

It turns out that our bodies have a set of three beautifully coordinated systems that help us respond to stress. The first part of this triad is our voluntary nervous system. This is the part of our nervous system that allows us to send commands to our body to get up and move. The basic parts of this system include what is called the primary motor cortex located in the frontal cortex and the pathway from that brain area via the spinal cord and nerves to all the voluntary muscles of our body. Voluntary muscles are the ones that we can move consciously and that can mobilize us to escape from dangerous situations.

The second key system that helps us respond to stress is that part of our nervous system called the autonomic nervous system. There are two parts of the autonomic nervous system that work in two very distinct situations. The first is called the sympathetic nervous system, responsible for our fight-or-flight response. When a stressor arrives in our life (lion, earthquake, nuclear disaster), it's the sympathetic nervous system that gets activated and prepares the body to respond. It does this by increasing heart rate, respiration rate, and pupil dilation (the better to see the charging lion with). The sympathetic nervous system will also send glucose into the bloodstream so the body and muscles in particular have quick access to energy and will also divert blood toward the major muscle groups in case we have to run. Other systems that are not needed in times of emergency are shut down, including kidney function, digestion, and reproduction. In other words, a lion attack is no time to pee, poop, or ovulate. Do it later.

The second branch of the autonomic nervous systems is called the parasympathetic nervous system, or the rest-and-digest system. This is the system that kicks in when we are relaxed, and basically works to reverse all the emergency 911 functions of the sympathetic nervous system. This system decreases heart rate, respiration rate, and pupil dilation. It sends blood and energy to the digestive system so you can digest that big Sunday brunch, supports reproductive function so women can ovulate and men can produce sperm, and allows for contraction of the bladder so you can urinate. The sympathetic and parasympathetic nervous systems coordinate their functions so that when one is active, the other takes the backseat and vice versa.

The third system involved in response to stress is called the neuroendocrine (hormone-release) system. This involves the secretion of two key hormones that are released in situations of stress to carry out some of the stress-response functions of the sympathetic nervous system. The first hormone is cortisol, made by the adrenal glands, which are located just above the kidneys. In response to stress, the sympathetic system signals the release of cortisol, which increases glycogenesis (the release of glucose into the bloodstream), suppresses immune functions, and decreases bone formation. In an emergency situation, that burst of cortisol helps activate our brains and our major senses so we are more alert to better deal with an emergency situation like finding the way out of a burning building. The second key hormone released during stress is adrenaline, also produced in the adrenal glands. It's the release of adrenaline in situations of stress that increases your heart rate to get the blood pumping; it also increases blood pressure, expands your air passages, and dilates your pupils. It's adrenaline that gets your body ready to run from that lion.

These systems have been beautifully calibrated to help in two major kinds of situations. The first is unexpected emergency situations analogous to a lion attack in the wild where there is an acute danger and your body's response systems get immediately activated and aroused for action. This system also deals well with non-life-threatening but short-lasting stress, as when you need a rush of energy to finish running a race, to get a big project done by deadline, or to make it on time to pick up your kids. This same system gives you the necessary burst of energy to get the job done.

THE STRESS SYSTEM'S DIRTY LITTLE SECRET

But as we have evolved and as our environments have become more complex, developing into complicated social systems, the sources of our stress have changed. In our plugged-in, online, 24/7 society, stress comes at us from all directions—from the guy talking really loudly on his cell phone on the train to the demanding boss to the competition to get ahead in your field. These are not fast-acting stressors. On the contrary, they are chronic, pervasive forms of stress. Note that these kinds of chronic, mainly psychological stressors were simply not present as humans were evolving in the plains and forests of Africa. The dirty little secret is that despite its sophistication, our stress system can't tell the difference between real life-or-death emergency situations and today's chronic psychological forms of stress. As a consequence, worry over paying your taxes can activate your stress system in the same way as a herd of charging wildebeests. Your worrying about taxes probably activates the system less than an unexpected wildebeest, but it gets activated all the same. The same is true of our perception of any event, circumstance, or troubled relationship: If we think of it as stressful, we experience it as stressful. It doesn't seem right, but that's how it works. If the sympathetic system stays active for these chronic stresses of our lives, then the parasympathetic system is never active, and your body and brain do not get any relief from the state of constantly being ready to flee or fight danger.

With chronic activation of the sympathetic system, you get all your emergency systems active all the time: Your heart rate is a little higher, your blood pressure is always elevated, and your blood glucose level is at a constant high, making less blood available for digestion and reproduction. It is easy to see how chronic sympathetic nervous system activation can lead to heart disease, diabetes, ulcers, and long-term reproductive problems such as erectile dysfunction and disruption of menstrual cycles. Not only that, but long-term stress weakens our immune systems, making us more susceptible to disease and prolonging recovery from injury. So while our built-in stress-response system is beautifully adapted to react to unforeseen acute dangers, it turns on us when chronic stress invades our lives.

And the bad news gets even worse. Long-term chronic stress negatively affects the brain. A long and rich history in neuroscience research has focused on these negative effects of long-term stress and, in particular, high levels of cortisol on brain function, and the story is not good. The three major brain areas affected by long-term stress are the hippocampus, the prefrontal cortex and the amygdala, which are the centers for memory, executive function, and managing emotion. Sound important? You bet.

The hippocampus is particularly vulnerable to stress because hippocampal cells are endowed with the largest number of cortisol receptors in the brain. A receptor is like a specialized doorway into a cell that allows particular hormones or neurotransmitters to modulate the inner working of the cell in a range of ways. As a consequence of all those cortisol receptors, hippocampal cells are highly responsive to any change in the body's cortisol levels. With a short exposure to cortisol, hippocampal cells work better and memory is enhanced (as in the pop quiz example I gave at the beginning of the chapter). But there is strong evidence that prolonged exposure to high levels of cortisol working through those cortisol receptors damages hippocampal cells and actually accelerates the aging process by damaging proteins and other metabolic machinery in brain cells. If you artificially increase the level of cortisol in the hippocampus in rodents, you impair the animals' physiological responses and cause shrinkage of the tree-branch-like dendrites (the input structures) of their hippocampal neurons. If the cortisol levels remain high for a long time, the hormone will actually start to kill hippocampal neurons, shrinking the overall size of the hippocampus. For this reason, long-term stress also significantly impairs long-term memory function. This is consistent with findings in humans who have endured long-term stress. For example patients with posttraumatic stress disorder (PTSD) or depression (specific conditions that are both strongly associated with long-term stress exposure) have significantly shrunken hippocampi and impaired learning and memory function, which suggest that long-term cortisol exposure has killed their hippocampal cells.

Many studies in rodents have shown that long-term stress also decreases normal hippocampal neurogenesis. When this happens, the normal infusion of new hippocampal cells starts to slow down with chronic stress. No more new brain cells! Stress will also decrease the synthesis of the growth hormone BDNF. Because BDNF is critical for the growth and maturation of the new hippocampal cells, less BDNF also means decreased survival of any hippocampal cells that do manage to be born.

Although the hippocampus has the largest number of cortisol receptors in the brain, the prefrontal cortex is highly sensitive to even short bursts of stress. As I've mentioned before, the prefrontal cortex, situated just behind the forehead, is essential for some of our highest-order cognitive abilities, including working memory (defined as the memory we use to keep things in mind, also referred to scratchpad memory), decision making and planning, and flexible thinking. Studies in animals have shown even relatively mild stress can impair performance of working memory tasks that depend on the frontal lobe. Physiology studies have shown that stress not only impairs the physiological responses and functions of the prefrontal cortex but also can quickly start to damage the dendritic branches of those cells.

The third key brain area affected by long-term stress is the amygdala, which is important for emotion and, in particular, for learning about aversive stimuli. Unlike the hippocampus and prefrontal cortex that are damaged with stress, increased stress in PTSD patients works to put the amygdala into overdrive such that scientists see increased amygdala activation in PTSD patients along with inhibited prefrontal function. In particular, the region of the prefrontal cortex that exerts inhibitory control over the amygdala is the region that has been shown to be underactive in PTSD. What does this mean? That PTSD makes people more easily aroused or reactive and impairs their executive functioning, including working memory and the ability to manage emotions.

WHAT STRESSES US OUT

Like everybody else, I've experienced a wide range of different kinds of stress throughout my life. Early on there was just baby stress. For me it was the end of summer vacation stress that happened every year. My poor mother would have to contend with a crying child who she knew enjoyed school but who could not let go of summertime fun. The stress escalated with the SATs, college finals, and graduate school applications. However, it wasn't until those last six months of my doctoral dissertation, leading up to my actual dissertation defense that I truly experienced my first major stressful event. This was hard. It was the moment of truth. After five and a half years of data gathering this was the moment when I had to synthesize all that data and finally see if I had something coherent, intelligent, and deep to say about all my work. The writing shouldn't have been so difficult—I knew what all the results were—but it was those pressures to be brilliant that made the task extra hard, that made me worry my conclusions would not be as earth shattering as I hoped that they would be.

To make matters worse, I had descended into a fast-food habit. To be specific, I became a huge fan of the Jack in the Box curly fries and a burger for dinner and there was no way I was going to make time to exercise. The icing on top of my homemade stress cake was that all the stress I was putting on myself was making it very difficult to sleep. I was staying up late every night to do as much as I could, and I was exhausted, yet I never felt like I got a good night's sleep because just as I was laying my head down on the pillow, the thought would pop into my head, Did I write enough today? That question would immediately lead to, Was the chapter I wrote good enough? and then, Will I be able to get enough done tomorrow?

Of course, you don't have to write a dissertation to experience this kind of stress. We all experience it: losing one's job, getting a divorce, having a health scare, worrying about money. These stressful experiences upset not only our daily routines but our ability to bounce back.

You will recognize that these are all examples of psychological stress. One of the worst things about psychological stress is that because it represents worry about what *could* happen, anything at all could become a form of psychological stress. And it never ends—it can become a vicious circle, as it did for me during that last half year of finishing my dissertation.

Situations that cause us psychological stress typically have four major characteristics. Psychological stress develops in situations in which you feel you have no control. Check! In my case, my fate was to be determined by my dissertation committee, and I felt I had no control over their judgments. It also develops in situations in which you have little or no predictive information; in other words, when we are faced with a big unknown. Check again! I received the okay to write my dissertation, but for much of the time of writing it, I had little feedback about its direction and had no idea of whether I was on the right path, leaving me to worry and wonder constantly about where I stood.

Psychological stress gets worse in situations in which we have no outlets for our stress. You may recall that at this time in my life, I barely had a social life. I had effectively taken my major outlets (hobbies, such as playing music) away from myself in the name of working hard. I did not give myself time to exercise or eat well or even sleep very well, and it would have been out of the question to spend time on frivolous pursuits, such as relaxing and going to the movies or dinner with friends.

The last situation that feeds the psychological stress monster is the feeling that things are getting worse. Even though I knew there was light at the end of the tunnel (that I *would* have to turn in my dissertation), I still couldn't get myself to believe that things were going to work out. I was so stuck in stress mode, it felt impossible to do the mental reality check that would have reassured me: *Wendy, it's going to be all right. You've worked years on this, you know your stuff.* No, this positive self-talk was not happening then.

Even if you haven't written a dissertation, I am sure you can relate to this story: a period of time when you felt besieged by stress. Maybe it's waiting for a house to be sold or preparing for a move or applying to school (for you or your child!). The good news is that we can use our brains to help ourselves manage these stressors. In fact, many stress-management strategies try to reverse or diminish the four major aspects of psychological stress. For example, in situations in which you feel you have no control, figure out where you *can* take control of the situation to enhance your feeling of personal power. If you feel you have no predictive information, try to ask more questions to get the information you need to address the problem. Are you worried about how you are doing at work, or do you think your boss or colleagues are speaking badly about you? Figure out a way to get some feedback so you can do a reality check. Taking control of even a small part of the situation can do wonders to relieve or at least diminish psychological stress.

Another big area of stress management revolves around enhancing your outlets for stress. Turn to your friends to help relieve stress or make new ones who will. Find hobbies that you enjoy that can transport you to your own personal happy place. It could be cooking, eating, walking outside in nature, or spending time with your pet—whatever works for you. And it will not be a surprise to learn that many stress-management programs emphasize both regular exercise and meditation, which many studies have shown can decrease stress and enhance your mood and feelings of well-being (as you will see in Chapter 10).

An important key is to do the exercise and meditation on a regular basis. You have to make it as much a part of your routine as the psychological stresses that you are battling for these stress-management techniques to be able to do the trick. The key is to find something that you enjoy (or you can learn to enjoy) as your stress-management solution, and you can decrease the stress and increase the happiness in your life.

SOURCES OF STRESS: RELATIONSHIPS

Maybe one of the most common chronic stressors in our lives (the kind that makes us sitting ducks) comes from difficult interactions with other people. Do you feel any anxiety around your parents or siblings? Does it get worse around the holidays? What about people at work? I had just such a stressful situation with a student in my lab where I didn't even realize how much stress it was causing in my life over many months until I was able to resolve it. When this student first joined the lab, there was an initial honeymoon phase when hopes and expectations were high on both sides. But then as our personality issues clashed (him: a mixture of frenetic and laid-back energy, me: thoroughly type A), those high hopes and expectations gradually withered away, and I ended up in denial about how bad the relationship was. He was smart, but from my perspective just didn't do the experiments I asked him to do within the time frames I asked him to do them in; he seemed to prefer to do things his own way and at his own pace. I found him petulant and generally unproductive, and I'm sure he found me demanding and overbearing. It got so bad that I simply tried to avoid the guy whenever I really didn't have to see him. I thought I was decreasing the stress with my brilliant avoidance/ denial strategy, but I was just making it worse.

Only when I was finally able to fix this broken relationship through another "doorman conversation" (see Chapter 5) did I realize just how stressful that daily interaction was on me. I remember I was even more nervous about this conversation than I had been about the one with my doorman because the last thing I wanted to do was make this work relationship even worse than it was. The thing that convinced me to actually have this conversation was my realization that as head of the lab, I was responsible for defining the kinds of relationships I wanted in my workplace. And this was definitely not one of them. I knew I needed to change it.

So I brought him into my office, sat him down, and said, "The fact is, our relationship has not been good for some time." Surprisingly it was a huge relief to admit that out loud.

I continued, " But I want to change it. My goal is make the rest of the time that you have in the lab (he had about eighteen more months of funding left), the best and most productive time that you have ever had. That's *my* goal, but I need to know from you what I can do to help make that shift. Can you tell what you would like me to change about the way I do things that would help you be more productive in the lab?"

Just like my doorman, he was stunned. I remember his mouth was hanging open in surprise. But to his great credit, he managed to collect his composure and say, "Well, one thing is, I feel like you don't give me enough credit for what I do in the lab." I thought about this and agreed with him. One of his positive traits was that he liked to help everyone else so much, which was part of the reason he didn't get what I asked him to do done. Instead of being grateful for his service to the lab and providing any kind word or acknowledgment, I remained annoyed at him because he was not getting his own work done. I told him he was absolutely correct and from now on I would be sure to acknowledge his contributions, and I did. There was a major change in our relationship after that conversation. Without even saying anything, he starting meeting his deadlines and making serious progress on this project. It was like another little miracle!

The amazing thing that happened was that not only did my relationship with that student change that day but the entire lab (a team of eight at the time) seemed to breathe a huge and simultaneous sigh of relief. I realized that our bad relationship had not only been stressful on the two of us but it had undeniably affected the entire lab. We all breathed easier after this conversation, and I eliminated a huge dose of daily stress that I knew was there but had no idea was so pervasive. After that conversation, it felt like there was more space, more light, and more laughter in the lab. Yes, stress is at its worst when you are not aware how deeply it's affecting you and those around you.

RESILIENCE

A large body of research has confirmed that long-term exposure to high levels of stress is associated with higher levels of mood, anxiety, and addiction disorders. But what about those shining examples of human resilience that survive horrific conditions relatively unscathed? People like Louis Zamperini, the former Olympian and prisoner of war survivor, and John McCain, who also survived long-term imprisonment as a prisoner of war? Studies done primarily in animals have started to reveal strategies and biological responses present in resilient individuals that appear to protect them from the terrible effects of stress. These responses include the following:

- Early life or adolescent exposure to chronic unpredictable stress helps buffer individuals from stress later in life. This phenomenon is called stress inoculation and suggests that there may be a critical period for stress exposure that helps build our antistress mechanisms. Thus there might be an optimum level of exposure for stress (typically moderate amounts of stress) at a young age that could be key to developing strong resilience as an adult.
- In animals, stress inoculation also increases the volume of a particular part of the prefrontal cortex (ventromedial prefrontal cortex—important for emotional regulation and decision making). This is the same part of the prefrontal cortex that has been shown to shrink in both humans and rodents that have experienced excessive stress.
- Studies in animals have identified a number of specific genes that are activated in resilient individuals in response to stress and antidepressants activate some of these same genes. This raises the possibility that we might find a wider range of ways to activate the resilience genes and better protect people against stress.

It's an exciting time for the neurobiology of resilience and this research in animals is providing exciting new directions for both treatment for and protection from the debilitating effects of stress.

HOW EXERCISE PROTECTS AGAINST STRESS

I have just told you that stress damages the hippocampus and prefrontal cortex and shrinks hippocampal volume by damaging dendrites, decreasing neurogenesis, and ultimately killing hippocampal cells. Given all we know about the positive, enhancing effects of exercise on the anatomy, physiology, and function of the hippocampus as well as the behavioral evidence from humans that exercise enhances attention functions dependent on the prefrontal cortex, it comes as no surprise to learn that research in rodents shows that exercise not only protects the hippocampus from future stressful situations but helps reverse the damage caused by long-term stress.

These studies are typically done by exposing animals to stress, exercise on a running wheel, or both and then assessing their responses on tasks designed to elicit stress/anxiety. I'll call it a rat stress test. For example, a number of studies have shown that rats given access to a running wheel for three to four weeks exhibit less overall anxiety than nonrunning rats on one of these rat stress tests. In other words, exercise seems to protect the rats from stressful situations that come their way. Other studies show that voluntary exercise seem to keep the rats calm and cool in a stressful situation in which nonexercised rats exhibited high levels of stress and anxiety behaviors (in rats, freezing is a typical measure of stress/anxiety).

But most of us are not sitting around waiting for stress to happen to us. We are currently in the middle of all kinds of stressful situations in our lives. What we really want to know is if we can help reverse the negative effects of ongoing stress. One of my favorite studies that addresses this question examines a particularly severe form of stress experienced by rat pups when they are separated from their mothers. When this happens, the pups exhibit stress-related responses, significant memory impairments, decreased hippocampal neurogenesis, and increased cell death in the hippocampus. However, if you put the maternally separated rat pups on a voluntary exercise regime after their separation and after they have already started to exhibit stress, you see the memory impairment go away, a reduction of the depressive behaviors, and a reinstatement of hippocampal neurogenesis.

But the key question here is, *Why* exactly is exercise relieving the detrimental effects of stress? One possible answer has to do with a theory that is gaining support called the adult neurogenesis theory of major depressive disorder. According to this theory, a decreased rate of neurogenesis is an important contributor to the depressed mood in patients with major depressive disorder. This idea is consistent with the findings that the size of the hippocampus in these patients and patients with PTSD is smaller than usual. This theory is also supported by the surprising finding that one of the initially unappreciated effects of some of the most common antidepressants is that they stimulate hippocampus neurogenesis. Not only that, but if you block the ability of the antidepressants to stimulate neurogenesis in experimental animals, the drugs don't improve mood. This means that the ability of antidepressants to stimulate neurogenesis is an important key to their effectiveness. It also shows the importance of hippocampal neurogenesis in regulating mood in general. This gives yet another perspective on the mood-boosting power of exercise. It's not enhancing mood only by enhancing levels of serotonin, noradrenaline, dopamine, and endorphins but also by stimulating adult neurogenesis. Incidentally, we now know that one of the previously unappreciated functions of serotonin is to stimulate neurogenesis in the hippocampus.

The adult neurogenesis theory of major depressive disorder helps us make some key links. For example, one key link is that the hippocampus is not just important for declarative learning and memory, as H.M. taught us, but also plays a key role in mood and is highly sensitive to stress. It makes sense that in rodents there is strong evidence that exercise improves memory function and decreases stress. In humans there is strong evidence that exercise decreases symptoms of depression, though the evidence for an improvement of declarative memory function is still quite weak. This means that with exercise you are getting two for one: With one activity (exercise) you get both stress reduction and cognitive improvement. And it seems to be doing both functions through the same mechanism: adult hippocampal neurogenesis.

This knowledge changed the way that I viewed exercise. At first it was something I needed for the general area of well-being in my life, something I should do if my mood and my time allowed. Now I think of it as an invaluable life tool, on the order of importance of my smartphone or my tablet. Just like my smartphone, I use exercise to make me smarter, more attentive to what I need to be attentive to, and to reduce the stress in my life. As I've described before, we still don't know all the answers to exactly how exercise is improving memory or attention or mood, but you can bet that I now know how I best benefit from it in my own life. If I'm too stressed, I take an exercise break. If I have a big talk or presentation coming up, I make sure I'm feeling good, rested, and well exercised. If I don't have time for a regular workout, I use one of my own exercise Brain Hacks or one of my favorite short workouts: the *New York Times* seven-minute workout. I now treat exercise like a tool to improve my life, and the more I learn about the neurobiology of the mechanisms at work, the more refined my prescription gets.

THE NUTS AND BOLTS OF MY NEW AND IMPROVED APPROACH TO STRESS

When I look back on how my current approach to stress developed, I have to say it started with my experience with physical exercise itself. When I was a stressed-out assistant professor working like mad to get tenure, it was exercise that first got me out of my head and got me to feel my body for the first time in a long time. It helped me regain my mind-body connection. In any exercise class, it's difficult to focus on anything other than getting through the workout. This automatically forces you to focus on the present moment. I see now that the exercise classes were not only getting my body in shape but giving me little bursts of present-moment awareness before I went back to work and started worrying about what deadline I needed to meet or how well I wrote that last e-mail. No wonder exercise felt so good to me. It's only in focusing in the present moment that you can really start to appreciate life as it is happening. I had too little of this in my own life and at first, these exercise sessions (which also often included a little meditation at the end) gave me my first regular taste of present-moment awareness. I have also used meditation to keep me focused on the present (see Chapter 10).

What is the key to mastering the stressful situations that I dealt with in my life? In retrospect, I believe that I was in the process of knowing myself better, knowing what I perceive as stressful (because stress is highly subjective), and becoming aware of how I deal with it or not. Part of what I was becoming aware of as I continued the process of rebalancing my life was that I no longer tolerated stagnant, chronically stressful relationships. I was actively seeking less stress and more joy in my life and was taking actions to make that intention a reality. My relationship with my doorman, with boyfriends who didn't work out, even my previously not-so-close relationship with my parents were all stressors that I was working hard to fix, one by one.

There was also a physical component of my strategy in dealing with stress. I was becoming more aware of exactly what was happening in my body as I react to stress. An understanding of the physiology and neuroscience of stress helped me change my relationship with stress and how I dealt with it. Back in those early days, I was thinking, worrying, and stressing over whether what I was doing was good enough or big enough or important enough to get me tenure and the respect of my peers. In fact, I believed that the level of stress and worry in my life was directly proportional to the value of the work I was doing. The most important people doing the most important things had the most stress and worry, right? Of course, I wanted to do highly valuable work so I worried and stressed over every detail to be sure everything would go exactly as I planned. In other words, I linked my level of stress and worry to my level of self-importance. I wasn't contributing something worthy or important unless I had a hefty and continuous amount of stress and worry in my life. At the core of this theory was my belief, which I described before, that I was only as good as the next paper I published or the next grant that I won. I published many papers and won millions of dollars in grants, but the stress and worry never ended because I was always working for the next one that might or might not be won. On top of this, I let my level of success as measured by papers and grants and invitations in my highly competitive field define my self-worth as a scientist and a person.

My regular exercise was showing me how powerful it could be to live in the present, focused not so much on mental worries but on what I was doing and how I was feeling, in my mind and in body. I started to become aware that my constant focus on future and past worries and especially on letting outside forces define my self-worth had to change. I realized I was letting my whole life slip by without ever being able to truly appreciate it because I was never in the here and now. How did I do that? I gradually started to shift my focus on all the outside criterion of success and importance (how many invitations to talks, how many invitations to write book chapters) to my own criterion of what made me happy, including a shift toward paying more attention to myself. This might sound strange because wasn't I getting tenure and publishing all those papers for me? Wasn't it my science reputation that was boosted with every paper and every grant? Yes, I did love science, I always had, but what got lost was an appreciation of my enjoyment in science rather than simply checking off a step on my way to becoming successful in science. I had started to neglect my own joy in these things and blocked off too many other avenues that could bring me joy, like a strong social network and art and music and laughter.

It was this shift toward a much deeper inner self-awareness and self-love that started to change my previous approach to stress forever. This is what gave me the desire to evaluate and the motivation to eliminate all the unnecessary forms of stress in my life that I could find. I want to be clear that for me this was not an all-out campaign to minimize stress. Instead, it was a powerfully focused intention to bring more joy, love, and happiness into my life. I was inundated by stress and, in fact, defined my success and importance by how much stress I had and could endure. What a shift to declare that I no longer wanted all that stress and instead wanted much more joy in my life!

It didn't happen overnight. I was battling forty years of theories and beliefs that kept a healthy amount of stress in my inbox at all times. But slowly and surely, I shifted my attitude about myself and toward myself and started throwing old sources of stress away. This does not mean I suddenly let go of all goals and deadlines. Instead, I became even more productive and energized because I focused more clearly on the goals that made me happy. For example, this allows me to say no much more easily if the request, however noble, does not align with my own life goals. This ability to say no without guilt has eliminated an enormous amount of stress in my life.

I always used to experience a lot of stress when I had to speak out in public. Not in lecture and prepared public-speaking situations, but in a town hall or faculty meeting where you had to fight to get your voice heard. While I was generally successful in speaking out in these kinds of situations, I was always worried about how my comments would be received or if my words might offend. This caused a disproportionately high level of stress and worry relative to their true importance. I still have a little jolt of adrenaline in these situations (the "good" kind of stress) but I have all but eliminated the more serious worry because I am clear about what I want and why I want it. I am far less concerned about how others perceive what I say and more focused on saying what I really mean in a clear and concise way.

As I got rid of more and more stress in my life, I found that I tolerated the unnecessary stress that remained less and less. I have described the difficult and awkward conversations I had, which essentially transformed my relationships with my parents, a doorman, and a student. Those three conversations helped shift my whole world from one where there was always a little stress oozing from the background—always hanging around the edges and hard to get rid of like that musty smell in your home after a flood—to one in which I could truly relax and let my parasympathetic nervous system kick in. I was able to have those conversations because I have become more tuned in to my own truth. In fact, the key to those conversations is simply being able to say your truth without anger or pride or ego getting in the way. I could have easily let my ego get in the way of asking my parents if I could say I love you to them. Why should I have to ask them? They are adults too, aren't they? But the truth was, I was the one who realized I needed to say it to them, so it was up to me to make the request. Similarly with my student, it was clear that I knew and he knew and every single person in my lab knew that we had a difficult relationship. I could have easily ignored it for the remaining months he had in the lab and simply blamed him for the whole mess. In fact, at an earlier time in my life, I could have easily seen myself doing this. But the truth was, I wanted a lab where everyone felt part of the team and felt respected, including me. I wanted to have a lab where all my students felt supported and could do good work, so instead of letting my annoyance or my ego get in the way, I simply told him how I wanted to be able to support him and asked him how he thought I could be able to do that better. I had to admit out loud that there was a problem and acknowledge that I was not only half of the problem but that I was fully responsible for solving the problem as head of the lab. That was a big hurdle to get over because it felt like I was admitting weakness. Instead it was simply acknowledging the truth.

TAKE-AWAYS: PROTECTING AGAINST STRESS

- The three biological systems available to combat stress are the voluntary nervous system; the autonomic nervous system, including the sympathetic (fight or flight) and the parasympathetic (rest and digest) subdivisions; and the neuroendocrine system.
- Too much chronic stress is toxic to both your body and your brain.
- The dirty little secret of the stress system is that it responds to and is activated by psychological stresses of modern life (say, high taxes and low salary) in the same way it responds to physical dangers (like a charging elephant).
- Long-term constant stress, including long-term psychological stress, has serious long-term health risks on cardiovascular function, digestion (ulcers), and reproduction.
- Long-term stress also affects widespread brain areas, including the hippocampus, the prefrontal cortex, and the amygdala.
- Moderate amounts of stress may help inoculate us and make us more resilient.
- According to the adult neurogenesis theory of major depressive disorder, exercise helps combat stress and depression by increasing adult hippocampal neurogenesis.

- The four major factors that cause psychological stress are (1) the feeling of having no control over a situation; (2) the feeling of having no predictive information about what might happen; (3) the state of having no social, leisure, or fun outlets; and (4) the feeling that your situation is only going to get worse.
- By reversing those four factors, you can decrease psychological stress on a situation-by-situation basis.

BRAIN HACKS: ANTISTRESSING *NOW*

Stress is an emotional response. Knowing this, we can interrupt its effects on our brains and bodies to lessen its impact. Here are some quick ways to combat stress.

- Hug or kiss someone you love. This could be an adult, a child, a baby, or a pet. Feeling the love can immediately combat even the most serious of stressful situations.
- Take some mindful alone time for yourself, such as a four-minute break for a quiet, relaxing cup of tea or coffee. Savor every drop of the drink and the alone time; make sure you don't look at your phone at all during this time.
- Call your funniest friend just to say hello.
- Dance by yourself or with a partner to your favorite song.
- Do any one of the exercise Brain Hacks from Chapters 5 and 6 (a jump rope or Hula-Hoop would be particularly good).
- Get or give yourself a hand or foot massage (this might take longer than four minutes but is one of my favorite treats).
- Write someone a thank-you note, e-mail, text, or Facebook message just for being them. Giving back can be one of the most powerful destressors around!
- Watch a viral talking animal video on YouTube. I love the one with the dog talking about bacon.

BRAIN HACKS: PSYCHOLOGICAL STRESS REDUCTION

These Brain Hacks will help you reorient yourself in situations that you perceive as stressful.

- Ask a friend you trust for suggestions on how to resolve your stressful situation.
- Go out for drinks with friends and don't think about your stressful situation at all for the whole night.
- Ask someone directly involved in the situation how she thinks it should be resolved. Let that open the conversation.
- Cultivate an attitude of optimism: if you don't have the solution to the situation now, one will come along.
- Seek out help from trusted supervisors, ombudsmen, therapists, or life coaches for solutions to big problems.
- Don't try to solve the issues all by yourself.

8

MAKING YOUR BRAIN SMILE

Your Brain's Reward System

As I did every typical weekday, I was standing on the subway platform about to board the number 6 train for work. But that day, I was not in a good mood. Why were there so many people on the platform? That woman just took the seat that I was headed for and didn't even look up—I *hate* when that happens!

Wait a second, why was I so grouchy? Was I hungry? Nope. Was I sleep deprived? No again. I realized that I was in such a foul mood because I had been traveling and it had been five days since I had been able to exercise. That was it! I was craving my regular exercise.

Since getting hooked on exercise, my body and brain now protest when I don't get my regular fix. As I explained in Chapter 4, we know exercise improves mood by increasing dopamine, serotonin, and endorphins in the brain. I look forward to that infusion of good mood, energy, power, and positivity that follows whenever I work out. The down side is that if I don't get my typical dose (four to six workouts a week on average), I start to feel annoyed and edgy; it's like something is bothering me that I can't identify. I go through what feels like exercise withdrawal. This response is what is typically referred to as a healthy addiction—something beneficial to your physical or mental health that you always make time for despite all the other obligations in your life. These activities are highly valued and sorely missed if something prevents them from happening. Yes, I have a healthy addiction to exercise. I also have a healthy addiction to massages.

It turns out that there are a lot of things in life in addition to exercise and massages that bring me great pleasure. They include (not surprisingly) delicious food, fresh cold watermelon juice, tickets to Broadway shows, river rafting, watching *The Sound of Music* with popcorn and hot chocolate, a surprising new finding from my lab, puppies, and Bach's solo cello suites (note that this list is illustrative, not comprehensive).

What is on your pleasure list? It turns out that every single item on my list has one characteristic in common. Each one activates the reward circuit in my brain. Our reward center is an evolutionarily ancient system that goes back two billion years and is crucial to our survival. Evolution has designed this system so that we find pleasure in those basic functions that allow us to survive and propagate: food, drink, and sex are central to that list. Rudimentary versions of this reward system are seen in worms and flies. These are called fundamental or core pleasures. But, of course, as living, breathing beings in a consumer-driven world we derive pleasure from a whole lot more than just food, drink, and sex. Our much more diverse list of pleasures is called higher-order pleasure. We derive delight from the people we love to spend time with; the places we go to relax and rejuvenate; and all that we spend money, effort, and time on.

Typically, what we value most in our lives can be found on these lists of pleasures. What's important to understand is that all of the key life decisions and choices that either bring more pleasure in our lives or limit it are influenced strongly by our brain's reward system. While one could argue that pleasure and happiness should be at the top of our wish list of things to understand about the brain, in reality only recently has serious effort been made to explore the neurobiology of pleasure and happiness. Fortunately or unfortunately, a lot of our current, nuanced understanding about the science of happiness, like a lot of neuroscience research, comes from studies of when this system is broken. In other words, we have learned the most about the brain's reward and pleasure systems from studies of addiction. In this chapter, I describe what we know about how the brain processes rewarding information for both fundamental and higher-order pleasurable stimuli as well as what we have learned about the brain's reward system from the study of addiction, including how exercise might help.

THE REWARD SYSTEM: A PRIMER

Before we get into the neurobiology of reward, it's important to define what we mean by *reward*. Reward is not a single process but a network made up of three distinct components. The first component is the one we most associate with reward and that is the hedonic pleasure component or liking. The second component is wanting, defined as the motivation for reward. The third process is learning, which includes the associations, representations, and predictions about past rewards that anticipate future rewards. The learning part of reward is carried out by two brain areas that we have already talked about a lot: the hippocampus and the amygdala. As we learned in Chapter 2, the hippocampus is important for making new associations and the amygdala lays down emotional memories, including those associated with highly pleasurable experiences. This introduction gives us a hint of how complex, interdependent, and interconnected the hippocampus and amygdala are as they contribute to many different kinds of brain computations.

What about the brain areas associated with liking and wanting? Studies done way back in the 1960s at McGill University by James Olds and Peter Milner (former husband of Brenda Milner of H.M. fame) were the first to identify what they called the pleasure or reward centers in the brain. This duo was looking for areas in the rat brain that, when stimulated, would inhibit the rat from doing whatever was associated with the stimulation. But as they were stimulating different brain areas, they instead found the opposite: brain areas that, when stimulated, would get the rat to keep doing whatever he was doing when the stimulation occurred. They found that if they allowed the rats to stimulate the electrodes implanted in these special brain areas themselves (the so-called self-stimulation experiments), the rats would obsessively stimulate the electrodes thousands of times and would forgo food to continue self-stimulating. These experiments first identified some of the brain areas important for reward and pleasure. The basic reward circuit includes a key brain area involved in perceiving and responding to reward stimuli called the ventral tegmental area (VTA). The VTA is located in the middle of the brain and contains the neurons that make the most important neurotransmitter for the experience of reward or pleasure: dopamine. The VTA-based dopamine-making (or dopaminergic) cells project to two important areas in the reward circuit: the nucleus accumbens as well as parts of the prefrontal cortex.

While those early studies by Olds and Milner were interpreted as identifying the pleasure centers in the brain, later work questioned whether these were centers for pleasure (liking) or desire (wanting). Dopamine release from the VTA is implicated in both functions. In fact, most recent work in this area of neuroscience has made progress in developing tasks and approaches to differentiate liking from wanting and suggests that these two states seem to use different parts of the same reward circuit in the brain.

How do you know if a part of the brain is associated with pleasure? First, you have to define when a stimulus is pleasurable. In
humans, it's easy, you just ask them. In animals, scientists use a
trick from Charles Darwin's book. Darwin did a famous study on
facial expressions and noted that all animals make facial gestures in
response to the environment, and we now know that many of gestures have been conserved across species, including facial responses
to pleasurable food, also called the yummy face. If you have seen
a baby eating food that she enjoys, you know immediately the face
that I mean. It turns out that you can identify that same face in rodents and ask if stimulation to particular pleasure centers enhance
the enjoyment of food (in particular sugary food) beyond how
they would typically experience it. It was found that stimulation
of two key brain areas enhanced a rat's liking of sweets. The first is
a particular part of the nucleus accumbens and the other is in the
ventral pallidum, a structure located deep in the forebrain (toward
the front of the brain). But these are not the only areas involved in
pleasure. fMRI studies in humans have identified a wide range of
cortical areas that are also activated during pleasurable experiences.
These include a part of the prefrontal cortex called the orbitofrontal
cortex, the medial portion of the prefrontal cortex, the cingulate
cortex, and the insula (which is deeply buried in the sides of the
brain between the temporal and frontal lobes). Other fMRI studies have shown that parts of the orbitofrontal cortex are activated
whenever subjects report the sensation of pleasantness associated
with chocolate milk. But once a lot of chocolate milk has been consumed, this area is no longer active and corresponds to a point when
the person reports no more pleasure from the treat.

A key open question is whether these pleasure areas are involved in simply encoding pleasure or if they are involved in causing the sensation of pleasure. The jury is still out. These areas are
clearly involved in encoding pleasure, but we are still working on

understanding exactly how the actual feeling of pleasure is generated in the brain.

While pleasure (somewhat sadly) is relatively understudied, the other side of the coin of reward, want, is extremely well studied in the form of addiction. In fact, we have learned the most about the workings of this part of the reward system through the study of this disease.

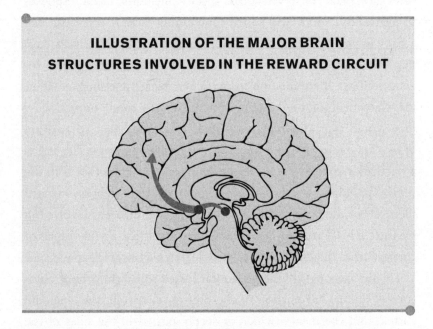

ILLUSTRATION OF THE MAJOR BRAIN STRUCTURES INVOLVED IN THE REWARD CIRCUIT

TAKE-AWAYS: REWARD AREAS INVOLVED IN PLEASURE/LIKING

- Reward includes liking (pleasure), wanting (motivation), and learning about future rewards based on past experience.
- Early studies by Olds and Milner identified specific brain areas that rats, when given an opportunity to self-stimulate, would obsessively stimulate for hours on end. This was our first insight into the reward system.

- The full reward circuit is a complex set of brain structures that include the ventral tegmental area, nucleus accumbens, ventral pallidum, several parts of the prefrontal cortex, cingulate cortex, and insula.
- The key subareas of the brain associated with pleasure are a specific region of the nucleus accumbens, ventral pallidum, orbitofrontal cortex, cingulate cortex, and insula.
- A key unanswered question is which of these brain areas or their interactions are actually causing the sensation of pleasure.

HOW ADDICTION MESSES WITH OUR BRAIN'S REWARD SYSTEM

The American Society of Addiction Medicine (ASAM) defines addiction in the following way:

Addiction is a primary, chronic disease of brain reward, motivation, memory and related circuitry. Dysfunction in these circuits leads to characteristic biological, psychological, social and spiritual manifestations. This is reflected in an individual *pathologically* pursuing reward and/or relief by substance use and other behaviors. Addiction is characterized by inability to consistently abstain, impairment in behavioral control, craving, diminished recognition of significant problems with one's behaviors and interpersonal relationships, and a dysfunctional emotional response. Like other chronic diseases, addiction often involves cycles of relapse and remission. Without treatment or engagement in recovery activities, addiction is progressive and can result in disability or premature death.

We know that release of dopamine is big part of our liking and wanting response. What drugs of abuse do, at least initially, is cause a much bigger dopamine hit (estimated to be two to ten times higher) than you typically get with a natural rewarding stimulus (for example, sex or chocolate), which is what helps make these drugs so intoxicating and irresistible. Many people (myself included) feel a form of addiction to exercise and negative withdrawal symptoms when we don't get our regular fix, but because the dopamine response is typically nowhere near as high as those from drugs of abuse, these responses don't come close to reaching the official ASAM definition of addiction. During the first stage of addiction, called acquisition, it's the artificially high surge of dopamine that can be the first step toward real dependency.

For example, cocaine acts directly in places where dopamine is released (such as in the nucleus accumbens) and blocks the normal reuptake of dopamine into the brain cells, resulting in a lot more dopamine floating around the brain than is usual. It's this high concentration of dopamine in the nucleus accumbens that causes the euphoric cocaine high. The normal brain is simply not used to that big a hit of dopamine, and the feeling you get is, as a consequence, unlike anything else you have ever felt. That's part of the reason it is so intoxicating. By contrast, heroin targets the brain's opioid receptors, which are found all over the reward circuit, including the VTA and the nucleus accumbens. Remember, receptors are the entry gates into the cells. Activation of the opioid receptors in the VTA stimulates the release of dopamine. Nicotine has yet another way to stimulate dopamine. When you smoke a cigarette, nicotine enters the bloodstream and activates receptors in the VTA called acetylcholine receptors, which in turn stimulate the release of dopamine. In this case, the dopamine release gives smokers that hit of pleasure with every puff. While all three of these addictive drugs give pleasure in the form of a high, each one provides a different kind of feeling because they are all activating the dopamine system in different ways, in different places in the circuit, and at different levels. It's those differences in the precise way the anatomical pathways are activated and the level of that activation that produce the different "flavors" of reward. Recent work has suggested that a major role of these drugs of abuse is to stimulate the wanting part of the reward cycle. While intense liking is certainly part of early drug acquisition, this system seems to quickly focus on the wanting part, which researchers are still trying to precisely identify.

After acquisition, escalation is the next phase of addiction, when drug use increases. One of the reasons escalation happens is because the very first hit of drug that spikes your dopamine feels amazing, just like the very first bite of ice cream on a hot summer day, but the fifth, sixth, and seventh bites don't feel the same; the only way to reclaim the initial feeling for a drug addict is to take more of the drug, more frequently. Over time your brain becomes less and less sensitive to the dopamine, and the decreased response drives you to take more and more of the drug to try to get back to that initial maximal dopamine response.

A major factor that helps determine predisposition to addiction is your genetic makeup. It is estimated that between 40 and 60 percent of a person's risk for addiction is genetic. You might think that people who become addicted get particularly intense levels of pleasure from the drugs, but paradoxically, they have a specific genetic modification that makes their dopamine receptors less responsive than people without that genetic modification. For people with a genetic propensity for addiction, a particular hit of dopamine (through alcohol, cocaine, sweet foods, or any other dopamine-stimulating substance) will cause less of a high than that same hit of dopamine in others. The people with a genetic predisposition for addiction will need six drinks to get to a level of intoxication that comes with two drinks for others; they need four packs of cigarettes a day, not just one.

Genetic factors contribute in yet another way to addiction factors. It turns out that drugs like cocaine affect the expression of many different genes within the nucleus accumbens, and one of those genes affected expresses a protein called DeltaFosB, which all of us have in our brain. With each injection of cocaine, you get a buildup of Delta-FosB in your nucleus accumbens cells that stays around for six to eight weeks, all the time building up more and more every time cocaine is ingested. There is evidence that the accumulation of DeltaFosB is the actual switch that activates addictive behavior. For example, just elevating levels of DeltaFosB in the nucleus accumbens alone, with no previous drug treatment, makes mice start to ingest more and more drugs relative to control mice. This is thought to be a molecular switch that keeps the addictive behavior going even when no drug is around. This is why those who have stopped abusing drugs often turn to other addictive behaviors; their neural pathways have been altered. This same protein also seems to be involved in the rewiring of the brain that occurs with long-term addiction. With long-term cocaine use, the dendrites (those branchlike input structures on neurons) in the nucleus accumbens become bigger and bushier. This in turn makes the neurons even more receptive to information from other areas, and scientists suspect that the areas that become more influential are the inputs from the hippocampus and the amygdala. This means that all the memories for the events, contexts, and emotions associated with taking drugs have an even stronger influence on the nucleus accumbens. This is thought to be the biological basis of craving: When memories of the drug-taking events get activated by the enhanced pathways and there is no dopamine around, the person experiences a craving. It's these long-term anatomical changes to the reward circuit that make recovery from addiction so difficult and relapse so easy.

While the vast majority of us will not become cocaine or heroin addicts, a different kind of addiction hits home for many more of us. It's called sugar. Many people feel addicted to sugar at one point or another. I felt it when I went through my Twix candy bar phase when I was training with Carrie (see Chapter 4). Like most things that bring us pleasure, sugar also activates the same reward circuitry as cocaine and heroin, albeit to a lesser degree. However, in one disturbing recent study researchers showed that rats given the choice between intensely sweet liquid and cocaine actually chose access to the intensely sweet liquid more than even high doses of cocaine, showing that in some situations, the sugar and sweet taste can be more rewarding than even cocaine. The scientists hypothesized that this striking effect may be due to the fact that mammals (including both rodents and humans) evolved in an environment low in sugar and, therefore, we may be hypersensitive to high concentrations of sugars. They hypothesize that exposure to lots of sweets, as is common in the modern world, might cause a hypersensitivity of the reward system to sugars, causing the response they saw in their rats. It's clear that an addiction to sugar is at least part of the problem in people with eating disorders, and scientists are starting to realize that sugar addiction can have serious consequences. We are still trying to understand the addictive qualities of sugar, how it relates to drugs of abuse, and how to cure that addiction when it happens. There are still no answers, but one promising avenue of research is the effects of exercise to curb addictive behaviors.

CAN EXERCISE CURB ADDICTION?

Some drug rehab centers are already strong believers in the power of exercise to treat addiction. For example, the Odyssey House, a

New York City–based facility for treating addiction, runs a highly regarded program that trains recovering addicts to run marathons. It's called the Run for Your Life program and was started by Odyssey House executive vice president, chief operating officer, and former drug addict himself John Tavolacci. Tavolacci credited marathon running with helping him beat his drug addiction. Odyssey House residents joke that before they sign up for the Run for Your Life program, the only running they did was away from the police. The program helps residents start slowly, with short but regular training runs through Central Park. They gradually build up to longer and longer runs, culminating in their ultimate event: the New York City Marathon. Talk about a runner's high! The Odyssey House believes in the power of exercise to help treat addiction. But what's the neuroscience behind this idea? The neuroscience is based on the interaction of exercise with the same reward system that drugs of addiction interfere with, and there is promising evidence that exercise intervenes at several key stages of addiction and becomes a replacement behavior.

First, strong evidence shows that adolescents involved in team sports or who exercise regularly are less likely than physically inactive teens to use cigarettes and illicit drugs. While these findings are suggestive, they don't prove that physical exercise actually causes the decreased drug use, they suggest only a correlation. However, research in animals does provide causal evidence that exercise decreases the chances of developing an addiction. In these studies, rats given the choice between a running wheel and the ability to self-administer methamphetamine will administer fewer drugs than the rats without the access to the running wheel. Similar results were found for alcohol. This suggests the very exciting idea that exercise can work as an effective substitute for drugs. While this level of exercise in rodents (or sports participation in high school students) clearly does not produce the same dopamine burst as the drugs themselves, it seems that it does produce enough of a dopamine buzz to compete with the drug consumption. While we know a lot about how exercise can work to decrease initiation of drug taking in rodents, more studies are needed in humans that directly examine the effects of exercise intervention programs on decreasing drug use. But progress is slow because the studies are both difficult and costly.

Another key phase of addiction that may be the most challenging is withdrawal from drug use. It is so challenging because it has been reported that up to 70 percent of recovering addicts relapse to drug use within one year after treatment. This is a period during which craving and depression can drive people back to drugs. The good news is that there is strong evidence in humans that exercise can have beneficial effects on withdrawal symptoms, particularly in smokers. Exercise has been shown to decrease cigarette cravings, withdrawal symptoms, and negative effect. The bad news is that nicotine is the only drug of abuse that has been studied thus far.

However, there are specific features of exercise that suggest that it would have a similarly positive effect during withdrawal from a much wider range of drugs. Specifically, all the data showing that exercise can decrease signs of depression and stress. Stress is a key trigger for relapse in recovering addicts, and as discussed, exercise works to decrease stress in myriad ways. With less stress comes less depression.

So, it is clear that exercise could be useful during the initiation, escalation, and recovery/withdrawal of addiction. All the neuroscience data suggest that this is because exercise uses many of the same pathways and activates the same reward centers of the brain as drugs, without a real addiction developing.

TAKE-AWAYS: BRAIN AREAS IMPORTANT FOR THE WANTING PART OF REWARD

- Drugs become addictive by overactivating the reward system that puts into motion immediate and very long-lasting genetic and anatomical changes in the reward circuits if the drug use continues.
- Because exercise activates the same reward circuits as addictive drugs it can help decrease the chances that drug use is initiated, it can curb the escalation of drug use if it has already started, and it can curb cravings in some cases and decrease stress levels that can decrease the chances of relapse.
- Exercise can become addictive itself and caution must be used when trying to design a program to replace drug addiction. It is important to monitor and minimize the shift to an exercise addiction.

ACTIVATING MY REWARD SYSTEM BY GIVING BACK

Life outside the lab was now just plain fun for me, with more and more social time and a growing circle of friends. We spent time eating, drinking, going to movies, and seeing shows together in the city. While I loved every minute of my new social life (I was finally starting to make up for my hermitlike early years at NYU), I found some of my greatest pleasures in activities in which I could give something back to the world. Since the summer of 2009 I have been teaching a free weekly and year-round exercise class for the NYU students, faculty, and staff. It's also free and open to anyone outside of NYU. At first the class was a great way to practice leading an exercise class right before the start of my "Can Exercise Change Your Brain?" class. But it became so much fun that I just kept on doing it. Many students in my "Can Exercise Change Your Brain?" class continued taking my weekly exercise class long after the semester was over, and I've met many students from all over the school whom I would have never met otherwise.

Some of my most memorable experiences from my weekly exercise class have come from the transformations that I have seen in my students. When I first started teaching the class, Pascale, a very bright, gregarious postdoctoral fellow in my department at NYU, was the only guy who came to class regularly. This did not bother him at all. Week after week, he took his place in the center of the front row and did the class with gusto. One morning, we both got on the elevator in our building at NYU, and Pascale asked if he could tell me something.

"Of course," I said.

He said, "You saved my life!"

I thought he was joking around. Or perhaps my lab had loaned him something he needed for his experiments. He went on to explain that when he started my classes he weighed close to fifty pounds more than he did now. Only then did I realize that the sweat shirt and pants he was wearing were hanging off him like a scarecrow. Then he showed me his old NYU ID, and the difference hit me like a ton of bricks. His face looked so much thinner and angular (in a healthy way) now! I don't know how I could have missed it. He went on to explain that he was in bad shape when he started taking class, in part because his work and school schedule meant he had not been exercising at all. He told me it was my regular class, together with the motivation of seeing one of the professors in his department teaching it each week, that kicked him into gear to exercise and start losing weight himself. He supplemented his classes with me with other exercises at home, but he credited my class with jump-starting his progress.

Pascale not only has kept the weight off but is keeping fit in new and innovative ways. He recently told me he had just gotten a treadmill for his desk at work so he could walk and work at the same time. He invited me to his office to check it out, and it was impressive. A beautiful shiny new treadmill that he had set up in front of his desk computer so he could keep walking during the many hours a day spent at his desk working, reading, and typing. For the last several years I have used a standing desk while I work to help decrease the inevitable slouching that happens when I type while sitting down, but I never considered adding a treadmill to my setup. Pascale has definitely inspired me!

My weekly exercise class is not the only way I give back. In fact, the reason I am familiar with the great work at the Odyssey House is that I organized a group of several exercise instructors to provide six months of free exercise classes for Odyssey House clients at the main Harlem branch. I loved getting to know my O House regulars, and it truly warmed my heart to see how grateful they were for the classes and all the teachers who donated their time.

Of course, I had to look into what was happening to my brain when I felt happy about giving back to the community. There are findings to explain the brain areas involved in those warm and fuzzy feelings that I always get when I'm giving back. Studies done at the University of Oregon measured brain activity in people given the opportunity to voluntarily donate to a charity. Many previous studies had shown that giving subjects money activates the brain's reward circuit. That makes sense! Who doesn't like getting money? What this study showed was surprising: When people voluntarily gave money to a charity on their own, it activated the same reward circuit as getting the money themselves. This is neuroscientific proof that giving is as rewarding as getting. In other words, generosity is rewarding and good for the brain. From a very personal perspective, I wholeheartedly agree with this finding. But it's not just blatant donation that I find rewarding. That very first time I stepped up and started teaching my fellow anatomy students what I knew about the structure of the liver, I remember getting a little jolt of pleasure. I thought it was pleasure from the act of teaching, but it was really pleasure from the act of giving. I think most teachers, and certainly all great teachers, are great because they are doing something that they love doing. Teaching is what is activating their reward systems, and it's the altruistic nature of the job that seems to be the key.

BRAIN HACKS: ALTRUISM

Try some of these four-minute Brain Hacks to activate your reward system through altruism.

- Pay the toll for the person behind you.
- Help a stranger in the street.
- Smile and greet someone you don't know in the street.
- Be kind to someone you dislike (extra points for this one!).
- Pick up trash on the street or on the beach.
- Tape your extra spare change to the jungle gym in the park for kids to find.
- Write a handwritten thank-you note to someone.
- Share your knowledge with someone.

LOVE, ROMANCE, AND THE REWARD SYSTEM

On the dating front, I was convinced that all the exciting new and sometimes altruistic adventures in my social life would soon attract equally exciting adventures in my romantic life. As I've mentioned, I am a firm believer in the idea that you attract into your life the kind of person that you are. I loved my life now, the person I was becoming and all the new friendships I had formed. I was ready and willing to start my next romantic relationship, and before too long someone new came into my life.

His name was Michael and we were introduced by a mutual friend.

The very first thing I noticed about Michael was his positive energy. He had a kinetic personality but also made you feel like you were the center of the universe when he was talking to you. He was funny and very sweet. The first time we met for a casual lunch, all I remember was how easy it was to talk to him. We chatted all through the meal and all the way out the door after lunch was over. When we finally reached that point where we had to walk in opposite directions, I remember thinking he actually looked sad to say good-bye to me!

It was the most endearing first-date moment that I have ever had.

For the next date we planned to meet up for a drink but I was (conveniently) starving so we went out for dinner instead. Despite his energy, Michael was quite shy at first, which was great because it provided a calm atmosphere as we got to know each other. We seemed to have lots of important things in common. We were both passionate about our jobs, mine in science and his in government, we both loved living in New York and had both spent time living in Washington, D.C. We also seemed to have the same family and life values. I particularly admired his close and loving relationship with his extended family. And he could really make me laugh.

Talk about activating the reward centers in the brain! I had this image of the dopamine neurons in my VTA firing like mad. In fact this image I had of my own VTA activation was accurate. Neuroscientists have started to study the parts of the brain activated in the early stages of intense romantic love, just like what I was experiencing, and it turns out that studies done in England, the United States, and China give surprisingly consistent results. All of these studies showed that the parts of the brain that become activated when subjects are looking at a picture of their beloved, compared to the activation seen when subjects are looking at a picture of an acquaintance, include the VTA and the caudate nucleus (also a major target of the VTA's projections). All studies agree that VTA activation represents the high reward value associated with seeing your sweetheart when you are in the throes of early romantic love. The caudate nucleus, like the VTA, has been associated with reward and motivation. For example, another study reported that when a monetary reward was predictable (in other words, if you find yourself a rigged slot machine that gives a payout every time you play), this same region of the caudate nucleus gets activated. So things that are a sure bet for high levels of reward activate the caudate nucleus. In addition to observing the consistent areas of activation when someone is looking at a photo of her beloved, researchers noted a consistent inhibition of the amygdala. The idea is that fear, processed by the amygdala, is decreased during periods of intense love. Based on my own personal experience, I would agree with this interpretation. These findings suggest that during the early stages of intense romantic love, your dopamine reward and motivation systems are working in overdrive and your fear response is inhibited. No wonder I felt so good!

In these studies, the authors describe intense, romantic love as: "Euphoria, intense focused attention on the preferred individual, obsessing about him or her, emotional dependency on and craving for emotional connections with this beloved and increased energy." They note that the combination of obsessive behavior with strong dopamine activation resembles key features of the early stages of . . . addiction.

I definitely had all the symptoms:

Obsessive about Michael and spending time with him—check!
Craving emotional attention from him—check!
Lots of energy—check!

Yes, I was definitely in that early addictionlike phase of intense romantic love.

But as I started to fall in love, I lazily started daydreaming about what it would be like to ride off in the sunset with my new prince charming and spend the rest of my days with him—to have and to hold till death do us part. Well, turns out that we also know about the brains of long-term happily married couples who report feelings of intense romantic love with their partners (lucky ducks!). Some of the same researchers who did one of the studies on early intense romantic love wanted to know if those same brain activations could still be observed twenty years into a relationship. What they discovered is that when one of the long-term lovebirds looked at a picture of her spouse the same brain areas were activated as those in people who are still in the early stages of an intense romantic love: the VTA and the caudate nucleus. But in addition, the researchers started to see other activated brain areas, such as the globus pallidus and the substantia nigra. These latter areas are interesting because they have been identified in studies that examine brain areas involved in maternal love. This suggests that long-term relationships activate the brain systems thought to be involved in social attachment. These brain areas have many receptors for two chemicals strongly implicated in attachment and pair bond formation—oxytocin and vasopressin. So such studies show that as a long-term romantic bond grows stronger, brain areas associated with deep personal attachment are activated. Now *that's* a pattern of brain activity that I aspire to!

This honeymoon phase with Michael was fantastic. Isn't it always? It was enhanced by romantic trips to Chicago, Miami, and San Francisco; long good-night phone calls when we were apart; and lots of intense time focused on each other when we were together.

We were not only in love but we were getting serious.

As our relationship progressed, inevitably, some differences started to emerge. They were challenging but not necessarily deal breakers. As you know, I am a foodie always in search of that great new restaurant. He lived for a great burger and fries (though he was in surprisingly good shape, given his preferred diet). I loved to go out with friends and meet new people—the more social invitations

I got, the happier I was. He felt socializing was a chore, didn't have a lot of close personal friends (bad sign), and would rather just spend time at home with me. I loved adventure travel and outdoor exploration of new cultures. He needed nice hotels and preferred to go on cruises. We managed fine through this, with me spending time with my friends on my own and us still spending time alone, though always compromising on where we would eat—Shake Shack (him) or Babbo (me), In-N-Out Burger (him) or the French Laundry (me). In the end we were always happy with a long binge-watching session of our favorite television series, something we almost always agreed on.

However, in the heady haze of love and reward responses swirling through my brain during those first six to nine months, there *was* one red flag that I probably should have paid more attention to early in the relationship.

From the beginning, Michael was very clear about the fact that he was separated but not yet divorced from his wife. He described the divorce as in progress and sure to happen sooner rather than later. I was more than happy to take him at his word. But as months and then a year and then a year and half went by, it became clear that nothing about this divorce was going to happen sooner; it was going to be later, an indeterminate amount of time later. It turned out to be way more complicated than I had ever imagined it would be. Michael maintained that the papers would be signed by this date and then by that date. Those dates went by so often with no progress at all, I had to start wondering if he was really getting divorced.

Then we started to fight about it. Regularly.

It became the focus of our relationship.

At that point, I could have just said, "You know what? I don't date separated men that aren't able to get a divorce." And I would have been done with it.

But when we were together he made me feel so loved. I just didn't want to give up.

Instead I said, "I need evidence of your commitment to me."

He said, "I can do that!" And about a week later we moved in together.

We experienced another honeymoon period.

I loved living with him. Or maybe I loved the idea of living with him. But the fact was that his road to divorce remained unrelentingly slow and was like the gorilla in the room. To be fair, I had no doubt that he wanted to get divorced, but his long-term inability to deliver on his never-ending promises to get the deed done started to wear on my trust in a serious way.

It turned out that moving in together was the beginning of the end for us. The broken promises and aborted deadlines just started eroding everything else. All those differences we noted at the beginning of our relationship (social engagement, food choices, preferred vacation destinations) that seemed manageable at the time now became unbearable. I wasn't going to be the other woman any longer.

I think at some point I just realized I was not in love with him anymore.

When I made that realization, I knew I had to break it off with Michael.

Despite all the clear signs that it was time for me to end the relationship, it was unbelievably hard to break up with him. In the beginning I had fallen deeply and passionately in love with him. I really thought we were going to get married. I felt a deep and terrible loss. But at the same time I knew it was the right thing to do.

I went through a very difficult time after we broke up. He had moved out of my apartment, but I still had plenty of reminders of him all around me. There were the knickknacks that he bought me as gifts and from his travels, the restaurants and stores that I passed where we had gone regularly, and even the times of day that he would always call me at work or at home. These were all the things that I had associated with that love that I felt for Michael. All the reminders, like the cues that cause craving and relapse in addicts, brought back a memory of my feelings. Maybe my love affair had been more like an addiction than I realized. In fact, all the reminders produced a deep longing in me. Not a longing to get back together with him, but a longing to feel that same intense romantic love that we had together.

My recovery from this breakup was long and slow. I continued to work out regularly, and I added more yoga to my exercise routine. I signed up at the last minute for a yoga retreat at a cute little inn called Good Commons in Vermont and had a great time. I met a bunch of interesting and yoga-minded people. I enjoyed the retreat so much I signed up again for a meditation retreat at the same place. These things together all started to help me feel happy and whole again. But it was a slow process. Honestly, I didn't feel completely recovered from the breakup for nearly a year.

As I was slowly but gradually healing, I realized that something else had changed for me. Namely, I finally (it was about time!) got clear about what I needed in a romantic relationship. First of all, there would be no more unavailable men of any kind. In a sense, Michael was just as unavailable as Daniel, the musician. Maybe I was seduced by the idea of seeing if I could win these unavailable men over to my side. All I knew was that I was not going to tolerate that kind of unavailability in my relationships from now on. That meant that I would no longer date married men, separated men, men who were too involved in their work for a real relationship, or men who were attached or unavailable in any other ways.

Michael also taught me an important lesson: Just because you fall deeply in love with someone doesn't mean it's going to work out. You have to know clearly what you need to be happy in a relationship and be ready to walk away—for the good of all concerned—if those elements are not there for both of you.

In addiction we know that the connection between the prefrontal cortex and the rest of the reward circuit becomes impaired, and this prevents the prefrontal cortex from using its decision-making powers to put the brakes on risky behavior. I suspect that while romantic love is stimulating the release of dopamine in our reward centers, it is also impairing our decision-making and evaluative abilities because we are so hooked on that feeling of love. At least that's how I felt. I could have used some help from my own prefrontal cortex during my relationship with Michael, but instead I ignored some pretty clear signs and made choices that I thought would keep what love there was left for as long as possible.

Well, we live and learn.

TAKE-AWAYS: CHARITY, GENEROSITY, AND LOVE

- Acts of true charity and generosity can powerfully activate the brain's reward system.
- Activation of the reward system may underlie the warm fuzzy feeling accompanying the act of giving to a good cause.
- Early intense romantic love can also powerfully activate the brain's reward system as well as stimulate the release of oxytocin and vasopressin, brain chemicals with links to social bonding.
- Intense love may have commonalities with aspects of addiction, including the obsessive behavior.
- The strong ties formed with places, events, and items associated with your beloved may cause regret and longing if the relationship ends, similar to cravings in addiction.
- We can recover from even the worst breakups and retrain our brains to learn from our mistakes.

BRAIN HACKS: MORE IDEAS FOR STIMULATING THE PLEASURE CENTERS OF YOUR BRAIN

We all know the things that stimulate our own pleasure centers. These are the things we dream we are doing or experiencing on a Monday morning instead of work. Here are some examples from my own personal list.

- Eating a meal you love.
- Drinking great Bordeaux (just having a few sips, as opposed to the whole bottle, will make it even more pleasurable!).
- Making love.
- Getting a full body massage.
- Watching your favorite movie of all time.
- Watching an exciting sports match.
- Playing your favorite sport.
- Reading a great book that you can't put down.

9

THE CREATIVE BRAIN

Sparking Insight and Divergent Thinking

Twenty years ago—even ten years ago—I would never have thought of myself as a creative person. I was a science geek who acquired knowledge, trained my attention, built up my memory for facts and ideas, and deliberately and consciously analyzed information. None of these skills seemed remotely creative. Indeed, like most people, I thought that creativity was the exclusive realm of artists, musicians, dancers, actors, and other people who seemed to express themselves in recognizably artistic or creative ways. Sure, there were innovators and scientists from Albert Einstein to Thomas Edison, Steve Jobs to Mark Zuckerberg, who are deemed creative because of the sheer brilliance of their work. But in a general way, creative thinkers seemed to have some unaccountable, mysterious quality that most of us do not possess.

Over the past few years, that has all changed for me. Not only do I now think of myself as creative but I believe we all have the potential to be creative. In many ways, this entire book is a narrative about my journey to find my own personal creative process—beginning with the connection I first made between exercise and my own brain and continuing up to this moment. For the past few years, I have been diving into all sorts of foreign territory for me, breaking down the barriers that have constrained the very way I think to uncover and understand just what creative thinking is all about. Creativity and science now go hand in hand for me.

And being creative feels different now too. I know that I am at my creative best when I feel open to life and its possibilities, I make connections between ideas easily, and I feel more spontaneous in my thinking and uninhibited by what others think of my ideas. And this perspective has carried over into my work: My research over the past few years has been much more varied, original, and spontaneous. Ten years ago, if someone had told me that I would be a certified exercise instructor studying the effects of exercise in people, I would have laughed at them! Today I bring African-style drummers to my talks to help demonstrate intentional exercise to big crowds made up of hundreds of people. I've come a long way, baby!

So if I now see myself as creative, what has changed? Am I now thinking in a different way from the way I had for the first twenty years of my career as a scientist? The answer to these questions is twofold: In some ways, I have always been a creative thinker, even if I didn't look at myself that way. As a scientist, I ask questions and constantly strive to look at problems in new ways. Yet, in another way, I also believe that I have become more creative.

So what do we mean when someone is creative?

In this chapter, I share with you not only how I discovered and embraced my own creativity but how you might be able to discover and embrace yours as well.

THREE MYTHS OF CREATIVITY DEBUNKED

Before we get into the discussion of the neuroscience of creativity, I want to address three long-standing myths about creativity.

Myth 1: Creativity = Right Side of the Brain

This idea is all over the Internet and is often proclaimed by the media. People are classified as either a highly creative and intuitive right brain type or a cool, collected, and highly analytical left brain type. Well, I'm here to tell you that there is no truth to the idea that only one side of the brain participates in or is responsible for creativity. While it is clear that language is housed (in most people) on the left side of the brain, the most recent studies suggest people who use both sides of their brain more are the most creative. Next time someone says he is a right brain creative type, you can simply tell him that the newest neuroscience research suggests that widespread brain areas focused in the prefrontal cortex are involved in creativity, and both sides of the brain are actually used for creative pursuits.

Myth 2: Only Certain People Are Creative

You can strike the myth that only some people are creative off your list of excuses for why you have not come up with the perfect solution for your home storage crisis. Creativity is not a mysterious process available only to geniuses like Matisse or Marie Curie. Recent evidence suggests that creative thinking is just a variant of regular everyday thinking and therefore can be studied like any other cognitive function. The difficulty becomes defining the best task or tasks to use when trying to study the brain basis of creativity.

Myth 3: All Creative Ideas Are Original

Despite my early dreams of being a neuroscience pioneer and discovering something that nobody else had ever even thought to look for, the truth is that the vast majority of creative ideas are based on preexisting notions. These new ideas are different, but are often built on the shoulders of previous work. This is especially true in science, where detailed and deep knowledge of all current research is the basis on which new experiments and new research are done.

But this doesn't make new ideas any less creative. Remember that saying, There is nothing new under the sun? True words and valuable to remember. Many of the most "creative" breakthroughs are better understood as creative remixes. One of the most famous examples is Steve Jobs and the personal computer. Technically, Jobs didn't invent any of the elements of the personal computer, Xerox did that. What Jobs did was perfect the technological tools and package them for the home market, which ended up starting an empire. Another famous example is Thomas Edison. He didn't invent the light bulb, but he sure did construct six thousand trials of materials for the filament and perfected it so it could be used commercially.

The truth behind these myths should serve to make us all a little bit more optimistic, and excited to be creative. I for one feel better knowing that my whole brain, and not just the right side, has the capacity to contribute to my creative thinking. I also take comfort in the thought that creativity is not some mythic ability that comes out of thin air but is grounded in normal cognitive processes and inspired by current bodies of knowledge. In other words, everyone is capable of creativity, and moreover, like any other cognitive skill—math, speaking French, working crossword puzzles, or playing Candy Crush—the more you practice creativity, the better you become at being creative.

THE ULTIMATE BRAIN MYTH: YOU ONLY USE 10 PERCENT OF YOUR BRAIN

If there is only one new fact you take away from this book, let it be this. The idea that we use only 10 percent of our brain is 100 percent false. We know from functional imaging and other studies that we use all of the brain, maybe not all the time, but with all of the cognitive, brain-based tasks we do all day and every day, our entire brain is getting a workout. So why has this myth persisted for so long? The answer to that question can be described as a combination of plausibility and hope. If it were reality, it would mean that each of us would be in possession of an amazing well of potential in our brain, if only we tapped into it. It's a myth tailor-made for the self-help industry too. Luckily, we do have the potential to build, stretch, and enhance the 100 percent of our brain that we are using every day because of neuroplasticity.

THE MEANING OF CREATIVITY AND ITS DIFFERENT FLAVORS

Over the past ten years, progress has been made in the field of the study of human creativity, and there is a growing but by no means unanimous consensus about the precise definition of the term. One definition of *creativity* is "the ability to produce work that is both novel (i.e., original, unexpected) and appropriate (i.e., useful, adaptive concerning task constraints)." The definition used by most scientists is "the production of something both novel and useful." In other words, creativity is about developing new ideas to solve old problems. Some examples are Uber, Airbnb, and Spotify. Despite this simple definition, the expression of creativity is essentially as wide as our collective imagination and can be achieved in a vast number of ways.

Generally, creativity can be deliberate or spontaneous (the Aha! moment). Each one of these two major categories of creativity can be further characterized as coming from a cognitive point of view or an emotional one. Many science experiments are characterized as deliberate cognitive kinds of creativity. These are typically experiments that discover something important and new but are informed by a whole slew of previous related findings. For example, my own discoveries of the importance of the cortical areas surrounding the hippocampus; discovering the role of the perirhinal and parahippocampal cortices (see Chapter 2) for memory is a classic example of a deliberate cognitive form of creativity. These areas had simply been ignored, and it just took someone to apply some powerful experimental approaches to identify their critical role in memory. But not all science experiments are deliberate. Other scientific discoveries have been inspired by more spontaneous Aha! moments. One classic example comes from Otto Loewi, a Nobel Prize–winning physiologist who studied heart functions in frogs. The story goes that he had a dream one night in 1921 during which he visualized a simple yet elegant experiment that would definitively show whether communication between different brain cells occurred through electric or chemical signals. He sat up in bed and scribbled some notes, but to his dismay, when he arrived in the lab the next day, he found he could not read his notes or remember the dream. Luckily for the field of neuroscience, the dream came to him again the next night and instead of waiting until the morning he immediately went to the lab and did the experiment that night. And so he showed definitively that in addition to electrical signals the nervous system uses chemical signals to communicate. Why was this experiment so important? We call these chemical signals neurotransmitters, and once identified, our understanding— both medically and scientifically—of how the brain works became much clearer. Loewi's discovery was an example of a spontaneous cognitive form of creativity. Isaac Newton and his understanding of gravity by watching an apple fall from a tree is another example of a spontaneous cognitive form of creativity.

What about the emotional side of creativity? This typically does not come out as often in the realm of science, but examples of emotional forms of creativity (deliberate emotional or spontaneous emotional) abound in the arts. For example, an example of deliberate emotional creativity are the cutout forms Matisse created, deliberately experimenting with different shapes and sizes and colors inspired by the emotional response he had to the resulting striking visual images. An example of a spontaneous emotional form of creativity is Picasso's famous painting *Guernica,* which was said to be inspired by learning of the tragic bombing of the city of Guernica in the Basque country of Spain during the Spanish Civil War. While these categories of creativity can be useful, creativity can also mean conceptualizing a whole new way to paint or sing or perform and then having the raw talent to execute that new conceptualization. Think of Frida Kahlo, Billie Holiday, and Lady Gaga—all of these artists reinvented their specific art form so that we experience a completely unique vision of the world through their painting, voice, and performance.

THE NEUROANATOMY OF CREATIVITY

Given the complexity of creativity and the wide range of different kinds of creativity currently recognized (deliberate, spontaneous, cognitive, and emotional), it makes sense that multiple areas of the brain are involved in these processes. One of the major brain areas involved in creative pursuit is the prefrontal cortex, a region that we've discussed throughout this book. Specifically, researchers have discovered that one particular subdivision of the prefrontal cortex, called the dorsolateral portion (DLPFC), is involved in three key functions critical for creativity. The first is working memory, which is the ability to process information online or, in other words, to keep information in mind as you are trying to solve a problem. Working memory is what allows us to monitor ongoing events and keep relevant information in mind so we can consider, evaluate, and mentally manipulate it to solve a problem.

Working memory is also involved in the second key function of the prefrontal cortex that plays into creativity: cognitive flexibility. Cognitive flexibility allows us to shift between modes of thinking and between different rules. Damage to the DLPFC reliably causes impairments in cognitive flexibility. Normal people can quickly and flexibly adapt to the changing rules, but patients with damage to the DLPFC instead perseverate (get mentally stuck) on a single rule and can't seem to explore other options despite feedback telling them that their answers are wrong. The ability to manipulate information in your working memory and then flexibly combine it, looking at it (in your mind's eye) forward, backward, upside down, and inside out personifies what creative people often do.

BRAIN HACKS: INVENTION

These Brain Hacks can make your brain think of a new way to approach a familiar habit or routine and possibly lead to creative problem solving and invention.

- Think of two new ideas to make your workday more efficient. You might rearrange your desk or the art on your walls. Or try changing the order in which you tackle tasks, starting with what you usually do in the middle of the day. Let this new order of activities or events create new neural patterns.
- Think of two ways to streamline the organization of your desk at work to improve productivity.
- Create a new kind of date to have with your partner, spouse, boyfriend, or girlfriend. Instead of going to your favorite restaurant, take an art, singing, or dancing class together. Or go to an interactive play or dance performance, in which the audience gets to be part of the action. Or try a brand-new kind of exercise class together.
- Try to cook something you have never cooked before, maybe something Persian or Russian or Cambodian. Try something that will allow you to play with new flavor combinations.

But that's not all the DLPFC does. This region has also been strongly implicated in directed attention, which is the ability to focus attention on a particular idea, item, or spatial location for long periods of time. This function is critical for deliberate forms of creativity in which attention to lots of things at the same time is often needed to sort through complex problems.

These three key functions of the prefrontal cortex—working memory, cognitive flexibility, and directed attention—are all critical for creativity. But this does not mean that this brain region is the sole site of creativity. Indeed, the prefrontal cortex is connected to other key brain areas that bring in and manipulate information in the service of creativity.

What are those other areas? As I mentioned earlier, emotion plays a critical role in creativity. Previous studies have shown a strong link between creativity and positive emotions such that people are more likely to have creativity breakthroughs if they report that they were happy the day before. Creativity is positively correlated with positive emotions, such as joy, love, and curiosity; art is often used to take viewers on an emotional journey through visual or auditory or even tactile stimulation. While emotions like fear and anger are typically not correlated with high levels of creativity, other studies show that strong negative emotional responses can sometimes be channeled into something positive and highly creative. For example, women whose children were killed in car accidents created Mothers Against Drunk Driving (MADD). This was a situation in which profound grief and anger sparked the creation of a powerful new organization. So there are examples of creativity being inspired by the whole spectrum of emotion from joy to devastation and back. Three key brain areas involved in emotional processing are the amygdala, within the temporal lobe; the cingulate cortex, located in the middle of the frontal lobe;

and the ventromedial prefrontal cortex, also part of the prefrontal cortex. The amygdala and cingulate cortex process emotional information and then send it to the ventromedial prefrontal cortex, an area involved in higher levels of social functions, personality, emotional planning, and emotional regulation.

THE ROLE OF IMAGINATION

There is still more to understanding what is happening in the creative process. Recent studies have shown that the hippocampus does more than just provide information in the form of long-term memory to the prefrontal cortex. It also seems to have a role in another important form of creativity: imagination.

Imagination is defined as "the faculty or action of forming new ideas, or images or concepts of external objects not present to the senses." Imagination is related to but not identical to creativity. While creative ideas can germinate and percolate in our imagination, imagination alone does not guarantee that these ideas are implemented. By contrast, creativity includes both the germination/percolation of ideas aided by imagination and the abilities that lead to implementation. In other words, it's fine to imagine something, but to follow through on that idea or insight is the truer measure of creativity.

The link between the hippocampus and imagination was first made by examining patients who had hippocampal damage. A research group in London examined a group of patients with damage thought to be limited to the hippocampus and a control group. The two groups were given tests in which they had to provide a description of new, imagined experiences. For example, the participants, none of whom had been to the tropics, were asked to imagine a scene in which they were lying on a white sandy beach along a beautiful tropical bay. One control subject responded like this:

It's very hot and the sun is beating down on me. The sand underneath me is almost unbearably hot. I can hear the sounds of small wavelets lapping on the beach. The sea is a gorgeous aquamarine color. Behind me is a row of palm trees and I can hear rustling every so often in the slight breeze. To my left, the beach curves around and becomes a point. And on the point there are a couple of buildings, wooden buildings.

A subject with hippocampal damage responded like this:

As for seeing I can't really apart from just sky. I can hear the sound of seagulls and the sea. I can feel the grains of sand between my fingers. I can hear one of those ship's hooters—that's about it.

Why would a brain area involved in creating the long-term memories for the events of our lives (that is, the episodic memories managed by the hippocampus) also be important for *imagining* events? These two functions may not be as distinct as they seem. The idea is that the same brain areas important for thinking about the past (such as the hippocampus retrieving memory) are similarly active when we think about the future (or using our imagination). Functional imaging studies have shown that a network of interconnected brain areas, including the hippocampus, is activated for what neuroscientists call past and future thinking. This is an exciting new development and suggests that the hippocampus is not just specialized in memory but also is involved in constructing episodes—both past and future. This ability also reinforces the hippocampus's ability to link items together as remembered experiences.

BRAIN HACKS: DIVERGENT THINKING

Sometimes creativity is thought of as divergent thinking: the ability to use an object for an original or novel purpose. Here are some exercises to help you think differently.

• Think of four new uses for common items that you see every day: toothbrush, toaster, stapler, rubber band, and so on.
• Think of a new way to drink your cup of coffee.
• Think of three new ways to ask your kids (or someone else) what they did today at school (or work).
• Think of three new ways to walk your dog or play with your cat.
• Find a new use for all the items that you would typically recycle.
• Find a new way to get to work and try to make it even more efficient than the way you are doing it now.

WHY STUDYING CREATIVITY IS SO COMPLICATED

While we are starting to appreciate how many different brain areas are involved in creativity, we are still far from a clear understanding of the neural basis of the creative process. One of the reasons creativity has been such a difficult nut to crack is the difficulty in finding a powerful and appropriate way to study the creative process. What is the best way to study creativity? Scientists have developed a few key tasks to be used as standard measures of creativity, focused on divergent thinking. One example is a test called the Alternative Uses Test. This test simply asks subjects to come up with all the possible uses for a brick that they can think of (for example, paperweight, doorstop, bug squisher, weapon). Most agree that this is a useful way to measure creativity, but it's important to keep in mind that this test alone cannot survey all aspects of human creativity.

One of the most powerful tools we have to study creativity is to observe patients with brain damage, just like the famous amnesic patient H.M., whom we discussed in the beginning of the book. One recent study tested a group of forty patients with various brain lesions and a group of control subjects. All of the subjects took the Alternative Uses Test. Scientists found that patients who had damage to the part of the frontal lobe toward the middle of the brain, particularly on the right side, had lower creativity scores, consistent with the old right brain equals creativity idea. On the other hand, they found that patients with lesions in areas on the left side, including in the parietal and temporal lobes, had higher than normal creativity scores compared to control subjects.

Wait a minute, *what?* Lesions on the left side can increase creativity? What's going on here? It turns out that brain damage leading to enhanced creativity has been seen before and is associated with a neurological condition called primary progressive aphasia (PPA). Damage to the language-associated areas on the left side of the brain, together with damage to the striatum, is commonly seen with PPA. Remember, the striatum is located deep in the middle of the brain and is associated with the brain's reward system but is also involved in movement. PPA is a degenerative neurological condition that gradually erodes speech and language functions. One of the most striking and well-documented cases of a patient with PPA is a woman named Anne Adams.

Adams received degrees in physics and chemistry before getting her doctorate in cell biology and worked for many years in academia. When she was forty-six, she took a leave of absence from her academic position to take care of her son, who had been in a serious car accident. During this time, she began to paint. Her early work was in the classical style, but relatively simple. Over the following six years, however, her painting evolved dramatically to a style that was bold, vibrant, and abstract, with great attention to detail.

When she was fifty-three, seven years before her first symptoms of PPA appeared, Adams painted what could probably be considered her masterpiece. She called it *Unraveling Bolero,* and she based it on the famous symphonic work *Bolero* by the composer Maurice Ravel. In this painting Adams meticulously translated Ravel's musical score into a visual modality.

Bolero is so powerful as a piece of music because it is unrelentingly repetitive, even perseverative in nature, building up to a surprising auditory climax at the end. Adams was incredibly systematic in her visual translation of the score. Each bar of the piece was represented by one upright rectangle, with the height of the rectangles representing the increasing volume of the music. The piece remains in the same key, represented by a unified color scheme in Adams's painting, until bar 3,236. At this point, Adams represents the dramatic conclusion of the piece with a brilliant burst of visually salient orange and pink bars.

Adams continued to paint voraciously, moving on to pieces focusing on abstract concepts, like the number pi (π), but then shifted her focus again away from multisensory and abstract themes to paintings focused on photographic realism. Initially, it seemed as if Adams were simply a woman who discovered her creative gifts later in life. But when Adams turned sixty, six years after she painted *Unraveling Bolero,* she began developing language problems and difficulty initiating speech. These were the first clinical signs of PPA. Sadly, her symptoms continued to progress, rendering her mute and affecting her motor functions; but through much of her disease, she retained the drive to paint and continued to do so as long as she was able to hold a brush. Adams died at the age of sixty-seven.

Because brain scans were taken on Adams from the time of her diagnosis until her death, we now have a rare window into not only the progression of her neurological disease, but also her creative output. MRI scans identified two key sites of change in Adams's brain. First, consistent with other patients with PPA, she showed severe damage in the left frontal lobe, which extended into the striatum, a subcortical region important for motor control. This left frontal damage included key language areas, which caused her initial language deficits. The striatum is the motor-related area damaged in Parkinson's disease and was likely the cause of Adams's difficulty in speech initiation. The damage on the left frontal lobe is consistent with studies, described earlier, suggesting that patients with damage to this region exhibit enhanced creativity. What else besides language do these left frontal regions do? It is thought that they supervise, or control, our attention and our ability to make particular responses. The idea is that if this part of the brain is damaged you lose your supervisory power over the attention and response areas, which could result in less supervised (less inhibited) and more creative thinking.

The second major change in Adams's brain was even more surprising. Researchers found that parts of her brain on the right side were actually significantly enlarged (compared to the brains of people of her age and education background). These enlarged brain areas included more posterior areas (toward the back of the brain) in the parietal and occipital lobes, which are critical in perception and imagery. This may have allowed Adams to make the links she did between the auditory modality of *Bolero* and the visual/perceptual aspects of her painting. In other words, it was no accident that Adams's art mixed two entirely different media—music and painting.

So what do we think was going on in Adams's brain? The idea is that her blossoming creativity in her fifties was initiated by the left side frontal damage caused by the earliest signs of PPA. This weakening of that brain area could have released the supervisory control over more posterior areas, enabling her to let her creativity flow.

It will never be known if Adams was born with larger parietal and occipital brain areas, or if their enlargement was a result of her disease. But it seems likely that those enhanced posterior areas played a role in her attention to visual and auditory detail and were responsible for the creative growth spurt that came later in her life.

Adams's case is striking in one other interesting way. While she was not aware of this when she painted *Unraveling Bolero*, Maurice Ravel composed his masterpiece when he was at about the same stage of PPA as Adams was. In fact, Ravel was probably the most famous patient with PPA to have been described in the medical literature. Ravel, like Adams, was attracted to repetition, which is a major theme in *Bolero*. But rather than being monotonous, Ravel created a growing tension in the piece with a beautiful and haunting melody that kind of carries us along, mesmerizing us until the very end. From her notes, it was clear that Adams was fascinated with Ravel's work. Her case, together with other examples of patients with PPA, suggests that one key to creativity is a release from control that may come naturally in some and from a neurological condition in others.

THE NEUROBIOLOGY OF IMPROVISATION

Adams's story provides another example of how information about patients with brain damage can help us understand brain function. But another approach to understanding the brain basis of creativity is to examine the brain activity in highly creative people. The tricky part is to pick the right category of artist to study. It becomes kind of like a riddle: Name an art form that can be quickly generated and evaluated, and can be performed while lying in an MRI scanning machine. Does such an art form exist?

My favorite answer to this riddle has been the study of musical improvisation, or the ability to quickly and extemporaneously create a melody. Two main forms of improvisation have been studied by neuroscientists: jazz improvisation on a piano and lyrical improvisation by rappers.

I have become particularly interested in the brain basis of rap. Several years ago I was doing a program for the World Science Festival in New York called Cool Jobs. The emcee of that event was a science rapper named Baba Brinkman. What, you may ask, is a science rapper? A science rapper is like any other rapper, but Brinkman raps about science. He's written rap based on the science of mating, evolution, and human nature. The two of us started talking about the neurobiology of improvisational rap, and I invited him to lecture at NYU on the history of the rhyme and rhythm of rap. His lecture naturally led to a fascinating discussion of the neurobiological study of the brain areas involved in improvisation and in rap.

While there is only one fMRI study that has examined the brain areas involved in improvisational rapping, there are more studies that have looked at the brain areas activated during jazz improvisation. In both the rap and jazz studies, scientists compared the patterns of brain activation during freeform improvisation with brain activity when the artists performed memorized pieces. The question was, What additional brain areas were activated in the improvisational condition compared to the memorized condition? In both cases, the results showed the same major patterns of activation within the frontal lobes. First, researchers found that in the improvisational condition there was activation of the part of the ventromedial prefrontal cortex on the left side. This region has also been associated with organizing internally motivated behaviors. In addition to the increased activation of this region, there was a deactivation of the dorsolateral portion of the prefrontal cortex in both the jazz and the rap study. The deactivated region is thought to be involved in self-monitoring and may be the origin of that inner critic that tells us, "Don't say that—that's stupid!" Or "If you do that, everyone will look at you funny." These self-monitoring areas appear to be inactive in situations of free improvisation.

Self-monitoring or, rather, the inhibition of self-monitoring is a critical aspect of all artistic performance and the creative process. It's fascinating that studies of improvisational jazz and rap have pinpointed the area that may be the key to letting go and going with the flow.

But the studies of improvisation are only in their baby stages. There are so many more fascinating questions to address, including what happens when rappers or jazz musicians start interacting with other musicians or with the audience and begin to respond to feedback. Are there structural differences in the brains of improvisational artists compared to the brains of other artists that might explain their talent in these areas? In the meantime, we have a small window on what might be happening in Jay-Z's brain as he performs live and off the cuff, as he is famous for doing.

FROM JAY-Z TO PHILIP SEYMOUR HOFFMAN: THE NEUROBIOLOGY OF ACTING

As I have told you, I grew up loving the theater and the movies. My childhood favorites included not just musicals but great dramas like *Gone with the Wind, Sophie's Choice,* and *The Godfather.* I very much admire actors who can make us feel as if real life is unfolding on the screen. I had an opportunity to get a deeper insight into the craft of acting at an event hosted by NYU's Emotional Brain Institute a few years ago called "Once More with Feeling." This was a panel discussion between the actors Tim Blake Nelson and the late, great Philip Seymour Hoffman and the neuroscientist Ray Dolan. The panel was moderated by actor-director and professor at NYU's Tisch School of the Arts Mark Wing-Davey. It was a fantastic event that started with some general questions for Hoffman and Nelson about their approaches to acting. Their answers were interesting, but the most memorable exchange of the night came when

Wing-Davey asked the neuroscientist, "Is acting akin to inducing a false memory? In other words, is it a true emotion up on stage or is it [something] different?"

Of course, no one really knows the actual answer to that question, but Dolan gamely offered an explanation. He said that acting was not the same as real emotion because, of course, when you are on stage, you are aware of the audience and you are monitoring your emotions in a different way from when you are feeling them for real. He suggested that there are some key elements of real emotion, but when acting, the emotions are just not the same.

No sooner had those words left Dolan's mouth than Hoffman immediately said, "I disagree!" He said that when he is acting, he feels every emotion he portrays.

When he said that, I think everyone else in the audience simultaneously thought, "And *that's* why you are such a brilliant Oscar-winning actor!"

Hoffman countered the argument that acting is different from real life because there is more monitoring happening. He said that we are always monitoring ourselves. We monitor ourselves when we go the grocery store to pick up some milk, when we are giving an important presentation in front of others, and when we are on stage.

He said, "The emotions that I have up there are real—even if the scene I'm playing is not. You are still living and experiencing life."

Nelson offered a different perspective, saying that when he has a fight with his pretend wife on stage it is different from when he has a fight with his real wife at home because the actor knows he is being watched when on stage.

But Hoffman stuck to his guns that people monitor themselves all the time, blurring the lines between real life and acting. He went so far to say, "I think people wake up and think, I should be paid to do this!" *He* certainly deserved to be paid to do it.

What became clear is that there are many different yet effective ways to approach the art of acting. Differences in philosophies of how to best play a scene abound, but all seemed to agree that when a scene is played well, everyone can appreciate it in the same way. That night I realized how difficult it would be to study the neurobiology of acting given all the different ways people think about the craft. My peers seem to agree, as I couldn't find any studies on this. The closest I came was an article in the UK paper *The Guardian* about an fMRI lab in London that studied the brain of actress Fiona Shaw. The study compared her reading lines from a poem to her simply counting a series of numbers. The story reported more activation in a region of the parietal lobe important for visualization when she was reading the poem. Unfortunately, this is where the findings end, and clearly this is a field left open for study.

BRINGING CREATIVITY HOME

While we may not be world-famous rappers or actors or have brain lesions that enhance our creativity, many of us (myself included) strive to maximize our creativity in everyday life. And luckily for us, neuroscientists and experts have useful information to help us spike our creativity.

Creativity gurus suggest that the concept of moderation is key to improving creativity. That is, while divergent thinking is useful for creativity, too much may lead to irrelevant ideas. Other studies have emphasized the importance of focused attention to enhance certain forms of creativity, but too much focused attention and you can lose the forest for the trees. Yet others suggest a shift in perspective or trying something counterintuitive can contribute to new insights, but a shift too large can take you too far away from the problem at hand.

So where does that leave us? Keep moderation in mind as you test your divergent thinking, shift your perspective, and focus your attention.

One of my favorite studies related to improving creativity was done by psychologists at Stanford University in 2014. This team studied the effectiveness of walking on creative thinking. I guess I am not the only one who had noticed that creative ideas suddenly come to me while walking the streets of New York. Stanford researchers tested this idea directly by comparing performance on a divergent thinking test (the Alternative Uses Test) during indoor walking on a treadmill and outdoor walking versus a control group. In one experiment, 81 percent of the participants increased their divergent thinking task score while walking relative to sitting. In another experiment, which focused on the generation of analogies, scientists found that relative to sitting, 100 percent of the subjects who walked outside generated at least one new high-quality analogy while only 50 percent of those seated inside did.

While these findings provide evidence that physical activity can increase aspects of creativity, the mechanism by which these effects occur is still unknown. It's possible that other forms of mild physical activity that free your mind, like knitting or fishing, would work in the same way. It's also possible that walking might improve mood and therefore make you more apt to be creative. So, while we don't yet know the connection between movement and creativity, the finding is both useful and immediately implementable. If you want a boost in your creativity or need to dislodge a creative block, take a walk. So, once again, exercise is good for your brain!

My own exploration of creativity has been one of transitions and evolution. I started out a very classic deliberate creative thinker, slowly building up my cognitive knowledge so I was able to ask interesting scientific questions about how memory works in the brain and how new long-term memories are born. I explored previously unexplored brain areas and discovered new things about them. More recently, I have moved beyond deliberate and focused creativity and started to explore the more spontaneous and emotional aspects of creativity in my work and in my life. My exercise research was inspired by my love of the practice and a genuine hope that I could harness the power of exercise to improve people's learning memory and cognition. I still use a great deal of incremental focused attention to study neuroscience and come up with key experiments that will help me reach this goal, but now I feel that my scientific work is infused with much more emotional resonance than it was when I started in this field. The artists, musicians, and other creative types in my circle of friends help feed the creative spark in me.

But maybe the biggest shift I see in my own creative life is my reaction to what my friend Julie Burstein, a public radio producer and bestselling author of the book *Spark: How Creativity Works,* calls the "tragic gap" or, in other words, the unknown. Early in my career that tragic gap terrified me. I knew this was part of science, but my approach to not knowing was just to put my head down and work as hard as I could until something interesting emerged. Maybe the strategy was okay, but the attitude definitely needed tweaking. The adventurous side of me was attracted to science for exactly this reason: to be able to explore the unknown corners of the brain and see what I could find. But then the reality of tenure and the number of high-quality publications I needed to produce to get tenure got in the way of that romantic dream, and I went about reaching my goals in the only way I knew how—with focused, unrelenting work.

The biggest change in my own approach to science today is that I am now able sit in that tragic gap of the unknown and appreciate that this is the place where the most creative ideas occur, when you don't know what the answer is or how an experiment is going to turn out. It's an uncomfortable, scary, lonely place, but ultimately, if you let yourself dwell there long enough it becomes, more often than not, a rewarding experience. The process involves letting go of expectations and quick answers and being open to strange ideas and intense feelings. This is where my meditation practice also helps enormously. And from this collision, new ideas emerge. No amount of controlling will help. You have to believe that by keeping an open mind and an open heart, you will encounter or discover an interesting path, and that seems, to me, the essence of the creative spirit.

There has been one last realization that I have made about my own creative process that I'd like to share.

While I'm convinced (and there is good supporting evidence available) that increased and sustained aerobic exercise has improved my learning, memory, attention, and mood, I also think that exercise may have improved my creativity. Why? Because exercise not only enhances the functions of the prefrontal cortex, which we know is important in creativity, but also enhances the function of the hippocampus, a key area involved in future thinking, or imagination. Improved mood has also been implicated in higher levels of creativity. Currently, this is not a proven fact, just a personal observation. But just as walking can help give you a creative burst, I suspect that long-term increases in aerobic exercise may work to grease the wheels of creativity and help you let go, be open to novelty, face your limitations, and sit happily in the tragic gap.

TAKE-AWAYS: CREATIVITY

- Creativity involves both sides of the brain and involves the dorsolateral prefrontal cortex interacting with emotional areas (the amygdala, anterior cingulate cortex, and ventromedial prefrontal cortex) and areas involved in long-term knowledge and memory (the cerebral cortex and hippocampus).
- The hippocampus is also implicated in imagination and future thinking, which are critical for the process of creativity.

- Some studies note that damage to the left frontal lobe causes a release from control, resulting in striking bursts of creative output in some patients.
- All these findings together support the idea that creative thinking is just a particular version of regular thinking that can be practiced and improved like any other cognitive skill.
- A key to creativity is learning to be in the tragic gap between ideas and to enjoy the process of discovering the unknown.

BRAIN HACKS: CREATIVITY

Creativity can be jump-started when you use more of your senses at one time. It can also be stimulated when you get out of your comfort zone and test your abilities. Try to learn something new!

- Use toothpicks and jelly beans or other soft candies to make a geometric sculpture.
- Get colored paper and try to make cutouts (like Matisse did) that look good to you.
- Go to the kitchen and create something good to eat using only what you have on hand (this will probably take more than four minutes, but you can try to plan the cooking strategy in four minutes).
- Make up new lyrics to one verse of a favorite song.
- Sit outside and blindfold yourself; for four minutes, listen to the world's sounds in a new way.
- Try to fix something in your house that you have never tried to fix before.
- If you are not already an actor, read part of a Shakespeare sonnet or a poem out loud with feeling.

10

MEDITATION AND THE BRAIN
Getting Still and Moving It Forward

L ike many of us, over the past several years, I have been inundated by information related to the benefits of meditation on the mind and on the body. Meditation can calm you down. Meditation can pump you up. Meditation can make you happy. Meditation can help you sleep better. Meditation can make you kinder and more altruistic. There have been many studies purported to suggest all the wonderful things meditation can do for us, but what if you just can't stick with it?

CONFESSIONS OF A YO-YO MEDITATOR

I came to meditation as a natural extension of my journey with exercise. As I have written, exercise changed my life. Mindful exercise, or exercise with purpose, created even more change and helped my brain to more fully respond to what exercise was doing. Meditation felt like the natural next step after intentional exercise, and with all the talk of the benefits, I was bent on making the practice a part of my life too. But I'll admit it: I'm a yo-yo meditator.

Even now, after years of making meditation a part of my life, I still don't have the kind of solid, unwavering, gotta-do-it-every-day kind of dedication that I want to have. I have never aspired to be a monk who could sit for many hours at a time in deep meditation. I was just hoping to get to a solid, reliable ten- or fifteen-minute meditation each day, but even that modest goal is easier said than done. For me (and I think for countless others), developing a regular personal meditation practice turns out to be a lot harder than I ever imagined.

It's not that I haven't tried. I am, in fact, a veteran of multiple meditation challenges. The first one was during my intenSati teacher training; we were asked to follow a twenty-minute YouTube video presented by Dr. Wayne Dyer called the *Morning AH Meditation,* every day. Dyer instructed us to meditate to the sound *Ah.* He explained that this sound is particularly powerful because it is included in the word for *God* in many different cultures: God, Allah, Buddha, Krishna, Jehovah. He also believes that the *Ah* sound is one of pure joy. According to Dyer there is something very powerful about vocalizing this sound while directing our attention to what we want to bring about or manifest in our life. The result, according to Dyer, is that when we regularly focus our attention through an *Ah* meditation practice, what we wish for starts to come true.

I was already a strong believer in the use of manifesting (another word for focusing my attention on specific things I wanted in my life), and I do believe mantras (like *Ah*) can help focus my attention during meditation. So I was happy to pair my intentions with the chants along with Dyer to try to kick-start my meditation practice. If the manifestations came true, so much the better!

At that time, my goal was to do the *Ah* meditation for thirty days straight, as we had been instructed in intenSati training, because doing anything for thirty days would make it habitual. I came out of the gates with a bang, and I definitely saw changes in my ability to meditate over that first month. In the beginning, when I sat down for my daily meditation, my right foot had the annoying habit of tapping as if it (or I) were impatient to get the meditation over with. I intentionally forced my foot to stop tapping and became much less fidgety in my month-long course. I also learned to control my breath better so I could sustain the *Ah* sounds during the entire meditation. To be honest, I got a tiny bit competitive trying to sustain my *Ah* as long as Dyer did (he gives a really long *Ahhhhhh* in this video). This was probably not the most Zen approach to meditation, but it helped keep me coming back.

Sure, I missed a day here and there, but I was pretty consistent for those thirty days. Did I notice any effects in my life during this exercise? Absolutely! I noticed that my focus was sharper; I felt less distracted and more efficient.

I was so pleased with the results, I gave myself a few days off as a reward, and before I knew it, I never started again.

Well, there went the idea that thirty days can set up a habit. I guess you really need thirty days *and* more motivation than I had at that time to create a daily meditation practice.

The good news was that even if I didn't continue with the daily meditations, they gave me a taste of what it was like to have a meditation practice, and more important, I noticed the benefits, including more focused attention and calm in my life. I knew I couldn't give up.

HOW DO YOU SUCCESSFULLY CREATE NEW HABITS?

I am a poster child for how to unsuccessfully start a new meditation practice. I failed so many times that it's a miracle that I ever managed to build the consistent routine that I have today. Why is it so difficult to start a new routine? Probably for the same reason that it's hard to start running regularly or reading every night before bed or eating more leafy greens. All of these are activities that take a certain amount of time, motivation, and even struggle. Learning something new is always a challenge. When I finally began to exercise in a regular way, it took a lot of mental, emotional, and physical energy to make that change. Yes, I was catapulted by the powerful epiphany about my own lack of fitness on my adventure trip in Peru and by my ever-widening presence in pictures from the time. But, ultimately, my motivation to really stick with it came from a combination of desires and positive outcomes: I wanted to feel strong, I wanted to lose weight, I wanted to be more social, and I began to see results. This positive reinforcement got me over the hump.

Other people develop motivation to change by completely immersing themselves in the new behavior. That's why many people go to boot camp for exercise or on meditation retreats. Immersion forces focus and regularity. It would probably have helped if I had gone to a nice long meditation retreat for which I paid someone to make me meditate every single day for many hours. That's also the premise of *The Biggest Loser* show or my personal guilty TV pleasure, *Extreme Makeover: Weight Loss Edition.* In *Extreme Makeover,* the producers remove the weight loss candidate from their home environment for at least three months to completely change their habits, while also transforming their home environment and filling it full of workout equipment to really make the new habits stick once the person gets back home. I didn't have either a major epiphany or the luxury of an immersive experience with meditation.

So what are you supposed to do if you don't have outside help? One useful strategy is to start much, much smaller than I did. For starters, a twenty-minutes-a-day commitment was just too long and resulted in immediate abandonment of the habit as soon as I finished my thirty days.

Instead, I would recommend starting with just thirty seconds a day, repeating one of your positive goals. BJ Fogg, a social psychologist at Stanford University and creator of a program called Tiny Habits, would add that you could pair your very short or "tiny" new habit with something that you already do every day. For example, pair your recitation of your intention with brushing your teeth in the morning. Just standing right there in the bathroom, close your eyes and recite the intention when you've finished brushing. Once you tackle this, you can build up to saying a mantra or doing a breathing meditation.

This idea was actually the inspiration for the four-minute Brain Hacks found throughout this book. You can easily turn your favorite Brain Hacks into a long-lasting habit by pairing them with your own personal daily anchor.

BRAIN HACKS: MEDITATION

Meditation is simple. It does not take long, it can be done anywhere, and it has a powerful effect on your brain–body connection. Try these quick tips.

- At the beginning of your day, take four minutes to recite one goal or intention for your life.
- Go to a quiet place outside and just sit silently for four minutes while focusing on the natural world around you and nothing else.
- Use a mantra like *Om* or *Ah* in a four-minute meditation.
- Before you go to bed, sit quietly for four minutes focusing on your breath.
- Find a meditation buddy and make a pact to do a partnered four-minute session together at least three times a week.
- Following the instructions given later in this chapter, do a four-minute loving kindness meditation just to start to get the hang of it. Rotate between breath meditations and loving kindness meditations to see which one you like best.

THE DALAI LAMA, AMBASSADOR OF MEDITATION AND NEUROSCIENCE

You might not know this but His Holiness the Dalai Lama is not only a global ambassador for meditation but also a powerful advocate for the neuroscientific study of the effects of meditation on the brain. I had the privilege of hearing the Dalai Lama speak on this topic in November 2005 at the annual Society for Neuroscience meeting. I was one of more than thirty thousand attendees at this meeting, and so I made sure to be at the session early enough to get a good seat. Miraculously, I was able to sit in the actual room where the Dalai Lama was speaking (many others were in satellite rooms with a video beamed in). The most striking thing about the Dalai Lama to me was his boyish charm. Maybe it was because he started his remarks by confessing, with a little giggle, that he had experienced stress in preparing his remarks for us neuroscientists. I just wanted to go up and pinch his cheeks. The Dalai Lama has an undeniable mix of joyful charm and profound presence, and both were on display that day in Washington, D.C.

He made the argument that there are many commonalities between Buddhism and the study of neuroscience. He said that the Buddhist tradition of exploring nature and testing ideas explicitly rather than relying solely on scriptures or unproven beliefs is consistent with scientific effort. Another interesting parallel between Buddhism and neuroscience is that the Buddhist tradition believes strongly in the potential for transforming the human mind. In fact, one of the major reasons meditative practices were developed as part of Buddhism was to help people change mentally, developing both deeper compassion and more profound wisdom. I was hooked.

In addition to being a worldwide ambassador for meditation, the Dalai Lama walks his talk; he meditates for four hours each morning. I can only assume that his rigorous practice came about through immersion because he was identified as a young child as the Dalai Lama and, therefore, started practicing meditation regularly and intensely at an early age. But I don't think you can explain his entire spiritual practice by the way he was raised. I believe there is something special about him. You can feel him as soon as you enter a room in which he is present—even one as big as a football field like the one I was in with him. He is a deeply spiritual and powerful presence.

The most surprising (and my favorite) part of the talk in Washington was when the Dalai Lama admitted that he finds meditation difficult. If the Dalai Lama can confess this, the rest of us should feel just a little bit better, right? He then challenged us neuroscientists: If we found a method for achieving the brain benefits of meditation without requiring four hours of practice each morning, he would happily use it.

BRAIN WAVES, THE BINDING PROBLEM, AND WHAT HAPPENS IN YOUR BRAIN WHEN YOU MEDITATE

Inspired in part by the Dalai Lama, in part by their own innate scientific curiosity, and in part by the growing interest in meditation's beneficial effects on a wide range of neurological states like depression and other affective disorders, neuroscientists have started to examine exactly what happens to our brains when we meditate. Is there a brain "signature" of meditation? And if neural activity actually changes in response to meditation, what does that suggest about our ability to control our minds?

One area of intense interest examines the patterns of rhythmic and synchronous brain activity, often called brain waves (the technical term is neural oscillations) that occur during meditation. These brain waves originate from the electrical signals of widespread networks of brain cells firing synchronously. The patterns of these waves of electrical activity can happen at different speeds from very slow, one to three times per second, for example, to very fast. Relative to meditation, some neuroscientists have been particularly interested in one of the fastest rhythms studied in the brain—the gamma wave, or gamma oscillation, which happens very fast, at about forty times each second. You can imagine the gamma oscillations like ocean waves of synchronous electrical activity that sweep over large swaths of the brain forty times per second and coordinate activity across these large brain areas. Neuroscientists have been particularly interested in gamma oscillations because previous studies have reported increases in gamma activity in different brain areas during a range of tasks of higher cognitive functions, like visual attention, working memory, learning, and conscious perception.

One well-known problem that gamma waves have been applied to is called the binding problem. The binding problem addresses the question of how the brain comes up with a coherent representation of individual items when so many different and widespread parts of the brain are involved in processing information from vision to emotion to smell to memory. But, as we all know, we don't see and feel the world in distinct snapshots, with vision separate from emotion separate from memory. Instead, our perceptions, thoughts, and actions are seamlessly integrated. To achieve this kind of seamless integration, neuroscientists have proposed that we need a master orchestrator to coordinate the perceptions, actions, emotions, and memories we process and that gamma waves could be just this kind of orchestrator. Scientists are currently testing the idea that gamma waves might help *bind* the activity of all the different brain areas responding to a particular stimulus (visual, olfactory, emotional), allowing our brain to create a coherent and integrated representation.

Because meditation was thought to change widespread brain states and improve attentional focus (remember attention had already been shown to be associated with increased gamma activity), it made sense to examine gamma wave activation during meditation. That's exactly what one group did. Their strategy was to examine the brain activity of expert meditators and see what might be different between their brains and the brains of control subject who had undergone just a week of basic meditation training. The expert meditators were eight Tibetan Buddhist monks who had accumulated between ten thousand and fifty thousand hours of meditation training over the course of fifteen to forty years. So they were truly experts when it came to meditating. The main experiment compared the pattern of brain waves, including gamma oscillations, as the monks and the novices performed an advanced form of meditation called the loving kindness meditation.

The neuroscientists wanted to see if there was any difference between the patterns of brain waves in the monks compared to those in the controls as both groups did the same meditation. The difference, it turned out, was like night and day. Of course, there were more gamma waves present in the brains of the monks relative to the novice meditators. In fact, the level of gamma activity was off the charts—the monks exhibited the most powerful gamma waves that had ever been seen in normal, nonpathological humans. This finding showed that there was indeed a dramatic difference between the brains of expert and novice meditators. These findings suggested that the extensive amount of training and meditation that the monks did resulted in the development of prominent gamma waves, which may reflect the monks' higher levels of awareness and mindfulness. But another possibility is that the monks were born with a propensity for more gamma waves that predisposed them to deep meditation, which in turn drove them to become monks. In other words, the brain waves of these monks might have started out different from birth. While this widely referenced study could not differentiate between these possibilities, other randomized control studies that compared the effects of meditation to no meditation in two equivalent groups of subjects have confirmed that the practice does cause a number of different kinds of brain changes.

OBJECT-BASED VERSUS OPEN MONITORING MEDITATION

It turns out that while there are many different kinds of meditation, the practice can generally be separated into two broad categories. The first category is often called focused attention meditation, which, as the name suggests, centers on the act of directing and sustaining your attention on a particular object. This is a very common form of meditation, often used in yoga classes when the instructor asks you to focus on your breath while pushing all other wandering thoughts away.

The second form of meditation is considered more advanced and is called open monitoring meditation. This practice can start with a focused attention meditation to calm the mind, but then the focus of attention moves from an object (that is, your breath) to a particular state, in the case of this study, to the state of loving kindness and compassion.

To start a loving kindness meditation, you first picture someone in your mind who you know and love or have known and loved in the past. It can work with an adult, but I find it works particularly well to picture a baby. Now, just savor the feeling of joy and totally uninhibited love that you feel toward that baby. If a human baby or a loved person in your life is not working for you, then puppies, kittens, or other baby animals might do the trick.

Once you have a feeling of loving kindness and compassion flowing through you, try to develop that same feeling toward others. You can start with the easy targets, your closest friends and family members. Then you practice directing feelings of loving kindness toward a stranger, like the person sitting next to you on a flight or the waiter who serves your meal. Then the hard part comes. You direct loving kindness toward someone you are having difficulty with or even hate. Of course, this last step could take months, years, or even a lifetime to master. That is okay.

A mantra that accompanies the loving kindness meditation is: "May you have happiness. May you be free from suffering. May you experience joy and ease."

You can offer these words to everyone in your focus during the loving kindness meditation. Good luck on this journey!

WHAT DOES MEDITATION REALLY DO FOR OUR BRAINS?

Although the gamma wave differences between the monks and the novice meditators are striking, this study did not provide specific information about the brain areas that might change with meditation. Gamma wave studies don't provide precise information about the brain areas sending the signals. fMRI studies have been done to help identify the areas and specific functions that change with meditation. One study compared brain activation in a group of meditators with at least three years of practice (not as expert as the Tibetan monks, but still pretty good) to nonmeditators. This study asked if there were differences in brain activity as subjects were performing a task of selective attention in which they had to quickly switch their focus from one thing to another. The surprising finding was that there was less activation in the frontal lobes during this task for the meditators than for the control group. This might seem counterintuitive at first, but it actually makes sense. If expert meditators have better control over their attention, they require less effort and, therefore, less activation to move their focus to different objects quickly.

But as I asked before, what if all these differences between the expert meditators and the novices were due to the fact that people who become expert meditators have brains that are wired differently from the rest of us (this is the same question that came up with the London cab drivers way back in Chapter 1)? To address this question, one study examined the effects of an intense meditation training (five hours a day for months) versus no such training in a group of volunteers between twenty-one and seventy years old who were all familiar with intensive meditation practice. These people were clearly not novice meditators, but it was a well-designed randomized control study nonetheless. Researchers randomly assigned the participants to either intense meditation or not to determine if the intense practice would improve attention and visual discrimination. This study clearly showed that the people in the intense meditation group had improved in both tasks relative to the control group. Another study with a similar randomized control design examined brain activity using fMRI as subjects focused on their breath, a very common meditation practice. In fact, the insula, located deep in the lateral (side) part of the brain, is known to be involved in attention toward internal bodily functions (for example, respiration or digestion). This study showed that meditation practice increased activation of the insular cortex relative to nonmeditators during a breathing practice. These are two examples of a growing body of studies suggesting that meditation training can change the brain activity of a random sample of people relative to no meditation training. But we will need many more similar studies to fully characterize the changes that occur with both short- and long-term meditation training.

Other studies focused on meditation alone have examined the effect of the practice not only on patterns of brain activation using fMRI but, just like the studies on exercise, also on brain size. For example, a number of studies have reported that different kinds of meditations result in increases in cortical volume (brain size increases). One study focused on people who had been practicing loving kindness meditation for at least five years. Researchers reported bigger volumes in the right angular gyrus and the posterior parahippocampal gyrus, brain areas associated with empathy, anxiety, and mood. Another study comparing expert meditators and nonmeditators found larger brain volumes in the right anterior insula, in the left inferotemporal gyrus, and in the right hippocampus.

Another study examined the effects of eight weeks of meditation in inexperienced meditators and found that, compared to before the program, participants showed increases in gray matter in the left hippocampus and posterior cingulate cortex.

You might wonder why all these studies are reporting different brain changes with meditation. One important note is that the studies used different forms of meditation, which makes direct comparisons across the research difficult to do. There are so many different kinds of meditation, and each kind could have its own set of unique brain effects. We have to be systematic about the kind of meditation that we study to start to sort this question out.

These challenges are similar to the impact of different forms of exercise. What is the difference between Nordic walking versus treadmill running versus spinning class on the results of scientific studies? Findings suggest that, although evidence shows that both exercise and meditation ultimately produce positive brain changes, there is still much work to do on specifying all the brain effects of the many different forms of exercise and meditation out there. We have our work cut out for us.

Taken together, all this research on the neurobiology of meditation tells us something quite extraordinary about the brain. We already know that aerobic exercise that changes so many physiological functions in our body—from heart rate to respiration to body temperature to muscle activity to the level of constriction and dilation of the blood vessels—can change the brain in striking ways. What the studies reviewed in this chapter show is that you don't have to move one finger to see brain plasticity at work. In fact all you have to do is sit very still and focus your mind, which results in significant changes in electrical activity, anatomy, and behavioral function. In a sense, that's even more surprising and more powerful than the changes seen with exercise. For me this is probably the most profound example of the great degree of plasticity that the brain is capable of.

AN EXERCISE MEDITATION SMACKDOWN

I am often asked whether exercise or meditation is better for your brain. What happens if you pit the raw physical power of aerobic exercise against the steely calm of meditation in a head-to-head, no-holds-barred, good old-fashioned smackdown? In fact, that was the question I kept asking myself. I had a strong exercise regimen and a growing meditation practice. I felt great, and I wanted to know why. What parts of my brain were changing because of exercise and what parts were benefiting from meditation? Was there one that was better for my brain?

The first important thing to note is that while we can study the effects of aerobic exercise on brain function in both animals and people, we don't have the luxury of studying meditation in animals—they don't meditate! Because of this fact, more is known about the detailed cellular and molecular mechanisms underlying exercise. Of course, in meditation's favor is the fact that it is a practice that has been around for thousands of years, and there is currently significant interest in this field.

Several recent studies out of Stanford have already compared the effects of exercise and meditation. One study compared a form of meditation called mindfulness-based stress reduction to aerobic exercise for improving mood in patients with social anxiety disorder (SAD), which is a common condition characterized by intense fear of evaluation and avoidance of social or performance situations. Using a randomized control design, the experiment showed that both meditation and exercise groups showed significantly fewer symptoms of SAD and better measures of well-being relative to the untreated control group. In other words, after both exercise and meditation, the subjects in these groups rated themselves higher in measures of well-being than did the people in the control group. In this study, both exercise and meditation appear to have similar effects on mood and well-being in subjects with SAD.

Although previous studies have shown that exercise improves attention function in elderly adults, a follow-up study showed that for people suffering from SAD mindfulness-based meditation may be more effective at improving attention than is exercise. In this study, researchers compared the effects of mindfulness-based meditation to aerobic exercise using fMRI to determine which parts of the brain became activated when participants tried to regulate their emotional responses about their own negative self-beliefs. The purposes of the study were (1) to determine which intervention (exercise or meditation) decreased participants' reactivity to statements of negative self-beliefs and (2) to observe the pattern of brain activity as participants responded to these negative statements. The researchers found that people with SAD who engaged in the mindfulness-based meditation showed less negative emotional response (as judged by themselves) while they were trying to regulate their reactions to their own negative self-beliefs than did the exercise group. In addition, the scientists noticed that the meditation group showed more brain activation of the parietal lobes (important for attentional regulation) than did the exercise group. The results suggest that meditation might enhance the attentional regions of the brain, allowing the individuals with SAD to better regulate their emotions toward negative self-beliefs. Although the participants in this study had SAD, all of us can benefit from the ability to better regulate our attention, particularly in emotional situations. We will need further studies to determine if the same benefit is also seen in normal control subjects who undergo meditation training.

PUTTING IT ALL TOGETHER

So who wins this exercise–meditation smackdown? Based on the data in humans that we have at the moment, it's a draw. There is evidence that both activities provide clear brain benefits. Both provide striking mood enhancement for both patient population groups and healthy control populations. Both can increase the size of various brain structures, and both have positive effects on attention.

I often ask myself what the best ways are to optimize the positive effects of exercise and meditation. Clearly, you want to be including both aerobic activity and mindful practice into your workout regime, which is exactly what I do. Maybe twice a week I'll take a Vinyasa yoga class, and another two or three times a week I'll go sweat it out in a spin, kickboxing, or dance class.

What does this all mean?

How many times a week should you do yoga? How many times a week should you meditate? How many times should you work out to get your heart rate up? What is the right number for optimal results? Is there a best time to practice? How long should each session last so that I can give both my body and my brain a chance to improve? These are the questions I ask myself.

We still don't know the final answers to these questions, but the neuroscience evidence shows us that both exercise and meditation have positive effects on the brain. For me, I feel best when I sprinkle all three throughout my week. I also pay close attention to how I'm feeling. You can experiment with your own concoction of exercises to try to find the perfect recipe for you. Yes, science should be able to make this all easy for us, and we are heading in that direction. In the meantime, the good news is that whatever combination and style and frequency of exercise and meditation that makes you feel the best is the one that works for you.

CONFESSIONS OF A YO-YO MEDITATOR, ROUND TWO: MEDITATION WITH PETER

About a year after my *Ah* meditation experiment, I was ready to try again. I signed up for Deepak Chopra's twenty-one-day meditation challenge. This is a free online program that sounded like a great way to reinvigorate my meditation practice. I jumped right in. I also told all my friends and the students in my weekly exercise class that

I was doing the challenge, so they could help motivate me. Despite all this support, I floundered pretty quickly, and I basically gave up on day three. I just didn't enjoy listening to what the leaders had to say, and there was someone new every day, so I couldn't get used to anyone either. I had failed Chopra's challenge.

Six or eight months later, I gave it another go. This time, I chose a different twenty-one-day meditation challenge, once again designed by Chopra, called "Manifesting Abundance." That title really appealed to me, and I signed on. Finally, something clicked. Maybe it was because for this challenge, Chopra had done all the meditations himself, and I just loved the sound of his soothing voice. I enjoyed listening to his stories too. Each meditation also had its own mantra, a Sanskrit word used to help focus our concentration. Chopra told us the meaning of each word and had us repeat the word throughout the meditation. One of my favorite of these mantras is "*Sat chit ananda*," which means "existence, consciousness, bliss." Another favorite, "*Om varunam namah*" means "My life is in harmony with cosmic law."

I think I was particularly drawn to these two mantras because of the sounds of the foreign words that I found beautiful to the ear and because of their meaning. Their sound and their meaning were both soothing and comforting, and I found myself able to focus on these mantras particularly well. It reminded me of when I discovered the intenSati class in the gym that really energized my physical workout. In the same way these particular mantras energized my meditation practice because I connected with them. So it seemed that mantras could actually work for me. I was thrilled when I completed the full twenty-one-day challenge! This was a first for me. I knew meditation had really clicked for me when I found myself going back to the beginning and starting over again. I was still a long way away from monk status, but I had definitely made a step in the right direction.

To tell the truth, the intention that I was focused on at this point was specific. I wanted to meet the man of my dreams, a partner with whom I could share my life. I had the exciting job, the great social life, and after recovering from the breakup with Michael, I was happy again. It was time to see what life would bring me next.

I knew I needed an intelligent, fit, social, energetic man who shared at least some of my love for food and travel and exploring New York. I also knew by then that I needed a man with a spiritual side, something that Michael had shown no interest in. As my meditation practice slowly grew stronger, I continued to keep that clear intention for a romantic partner in mind. And then, a curious thing happened. I met a man who liked to meditate! And he was really hot too!

One day, on a whim, I decided to try the online dating sites again for the first time since Michael. After many weeks of perusing the sites with little luck, I finally saw someone who looked really interesting. He had a charming profile; he seemed very intelligent and was clearly an athlete. And he loved to dance—double gold stars in my book because I love to dance and have never gone out with anyone who could really dance. I learned his name was Peter, and we arranged to meet for a drink at one of my favorite Italian wine bars in the city.

The first thing I noticed about him was that he looked just like his online picture: very handsome! The second thing I noticed about him was his very calming presence. I soon also learned that he had a strong meditation practice that he had developed over many years. We also discovered we had something else in common: We both spent a formative year in France, he as a high school student and me as a junior in college. We both spoke French fluently, but needed someone to speak with. By that point, my French was quite rusty, and speaking French with Peter that night forced me to dig deep to find that French vocabulary hidden somewhere in the recesses of my brain. In addition to speaking French beautifully, Peter was also musical; he liked to sing and fool around on the guitar. I think I've had a soft spot in my heart for musicians ever since François, so here was another thing in Peter's favor.

But of all the interests we had in common, the one I was most excited about was his meditation practice. Maybe it was because he had such a deep and calming presence. I wanted to learn more about him and his practice and bask in the glow of that presence as much as I possibly could.

Peter and I had a whirlwind romance. After that first date at the wine bar, he invited me out to dinner and dancing with live music. It was one of the most fun dates that I've ever had. We followed that date with one that included a dance performance and a group of my friends, then more causal dinners and evenings playing music and singing (well, he did most of the singing while I kind of hummed along).

Peter told me that he had started his spiritual journey when he went to a yoga retreat and found a teacher who really spoke to him. The teacher described spirituality as a way of life and not a religion, and Peter really responded to that. He had been following this teacher for years by the time we met, going to retreats with him, reading his books, and using his teachings in his everyday life as much as he could.

These discussions forced me to think about and then articulate my views on my own spiritual practice—and this was new to me. I had never been religious growing up, and as an adult who happened to have a job as a neuroscientist, it felt awkward to even consider any kind of spiritual practice. Meditation as an exercise in attention was just fine, but a common opinion of many of the quantitative and empirical minds around me during my science education and career was the harsh view that religion and spirituality were for the weak and feebleminded. Eventually, however, as I got deeper and deeper into my own meditation practice, I was clearly drawn to the spiritual side. In most organized meditation sessions, tapes, or other presentations, including yoga classes, references to the spirit or the universe or simply energy were always present. Consistent with this idea, I found there was a power in the act of meditation. For me, that power came from going within and the feeling that I was tapping into a kind of universal energy that linked me to all others. For me, this experience felt spiritual in nature. I would say that at this point, my spiritual practice is still in its early stages. I'm still learning and exploring. Because of these conversations with Peter, however, I had started to think more about how my spiritual practice relates to my life as a neuroscientist. No, I didn't have empirical evidence that there was a spirit or any special energy source linking us in the universe. I'm not sure if we actually have the ability to even do the kind of experiment to prove or disprove this idea. But what I did know was that I was committed to continue exploring the spirituality through meditation and see where it took me.

The Dalai Lama said that Buddhism shared with science the goal of exploring nature. I feel that my meditation practice is an exploration of both spirituality and my own true nature. I may not be able to prove the existence or inner workings of a larger universal energy, but I'm still interested in exploring how it works in my life. In a way, this is exactly how I approach the science questions that I'm tackling. I don't know exactly how exercise is affecting brain function, but I can develop my own hypotheses and take steps to test these hypotheses or ideas systematically. Similarly, I am developing my own hypotheses about how spirituality works in my own life, and while I'm not doing large-scale experiments to test these ideas in a randomized controlled fashion, I can and do test them in my own personal spiritual practice. At their heart, my pursuit of the understanding of how exercise affects the brain and my pursuit of an understanding of meditation and spirituality in my life share my own curiosity about the nature of life and experience.

After all this talk of meditation, Peter and I eventually started meditating together. While I enjoyed my solitary morning meditation practice, I also loved meditation in a big group too, like what happens at the end of a yoga class or at a meditation retreat. When I meditated with Peter, it didn't feel exactly as if I was in a whole roomful of meditators, but it was very different from meditating alone; he made the meditation somehow feel deeper just because he was there. It was also clear that nothing was going to interrupt his practice when he started meditating, and that focus transferred to me.

A couple of weeks after we started dating, Peter and I decided to play hooky, just because we could. We went skiing in Colorado. He was a beautiful skier, and despite a bout with severe altitude sickness that hit me in the middle of the first day, I bounced back and had more fun skiing with him than I had had in many years.

Our relationship was going so well. He was handsome, athletic, intelligent, spiritual, and a great dancer. What more could I ask for? We just got along in so many ways.

But there were ways that we didn't get along. His monklike demeanor was fantastic for a good meditation, but he could be very independent, maybe even verging on distant. Sometimes, we would have fantastic and fascinating conversations and other times it felt hard to find things to talk about at all. Not because he was not interesting but because he didn't always want to interact at the same level that I did. These instances left me feeling lonely and wanting more.

But more important, I realized quite quickly (after only a couple of months) that while we shared many interests and I appreciated many of his qualities, I didn't love Peter. I just really liked him. I asked myself what it meant to be wishing that a guy were different this early in the relationship. If I didn't really love him, and I wanted him to be different, I decided it might be time to move on.

So I told Peter exactly what I was feeling. He was very gracious. He said he was grateful for my honesty and my self-awareness.

I'm sure this was the right decision because Peter and I are still good friends today. In fact, I must admit that my breakup with Peter was the most adult breakup (yes, there is such a thing) that I have ever experienced in my life. And there is a reason for that. My meditation practice was starting to pay off. I found that meditation was beginning to shift the way I responded to the world in three very positive ways. The first was that I was becoming more self-aware and grounded. I was becoming clear on what I really valued and what I wanted out of life; to be fair, this realization was likely a combination of meditation and exercise plus life experience. But the quiet internal contemplation of meditation helped me develop this improved awareness. With this improved awareness came appreciation—appreciation of my life at all levels from awards or recognition to wonderful friends to all the things that bring me pleasure during the day, such as meditation, tea, a great workout, a walk through the city, or (one of my favorites) finding a new restaurant to try.

I had spent so much of my life using maximal effort and will-power to reach my goals. Initially it was my goal of building a great research lab and getting tenure, and then it was my goal to rebalance my life, starting with exercise. Meditation not only allowed me to finally appreciate all the benefits that all that rebalancing brought to my life but also allowed me to realize that maybe I didn't need to use so much effort to bring these changes about. This lesson took me quite a while to learn. It went against everything I had ever learned about how the world works. My old view: Work hard and at 100 percent for what you want because no one is going to help you. My new view: Enjoy what you have, and look for the signs telling you which direction to go next. And you will have help along the way.

The second way that meditation has shifted my worldview is that it has helped me live more in the present. As I described in Chapter 7, exercise first got me to appreciate focusing on the present moment, but it was only with more regular meditation that I really started to develop this skill. I had been living so much of my life focused on a future goal and doing everything I could to get that goal that I never gave myself time to live and appreciate the present moment. I now notice when people are really in the present moment. There is a presence to them that contrasts so strongly with the distracted, smartphone-checking interactions we have with so many people. The people living in the present moment are there, they are focused, and they are taking it all in. This is part of the reason there is something so striking about the Dalai Lama. He is hyperpresent in the moment, and we are just not used to seeing, feeling, or experiencing that. I'm certainly not present all the time, but I am much more sensitive to when I am and am not, and I am on my way to increasing the amount of time that I am living in the present moment.

The third way that meditation has impacted my life is probably the most important. It has been by bringing more compassion into my life. This started with the loving kindness meditation. I admit that first I thought that the goal was send more compassion out into the world. In other words, be nice. But I have come to realize that this is not the way it works at all. Instead, I realized that I needed to first focus the loving kindness meditation on myself. I found it surprisingly difficult to do this, even more difficult than the idea of being loving and compassionate to others. I realized that I had been so hard on myself for so long, that I wasn't working hard enough or wasn't successful enough, that one of the most profound things that the meditation practice brought was quiet time to consider my own loving compassion to myself. With meditation, I have learned how to love and trust myself to a level that I never had before. Yes, I am more confident and much happier. This is just because I have more friends and a more balanced life. The confidence and joy are also grounded in my own self-approval. This is huge for me, and many others, I suspect. At some point, I had to ask myself, Do I approve of me?

My answer to myself was, Am I *allowed* to approve of me? I had spent so much time using external measures of my own success that self-approval had somehow disappeared. The practice of meditation has allowed me to shift my practice of using only external sources to judge my self-worth and instead start using my own scale to measure how successful I am, how happy I am, and what it is I want to do next. Only when I really started loving and appreciating myself could I then turn that love and compassion to the world around me. Loving myself and others has shifted the way I see the world, which is no longer a teeming snarled snake pit to survive but a lovely garden to find my way through.

This may sound like I'm giving meditation too much credit for these personal changes, but keep in mind how many years had gone into transforming aspects of my life, from how I spent my time to how my body felt to cultivating my own mind-body connection. I was ready for the next shift that meditation brought about. I had primed my brain for plasticity, and the meditation seemed to bring it all home.

How has my behavior changed because of this new level of self-love? Focusing on loving and appreciating and trusting myself has taught me how to better love and appreciate and trust others. I know now that even something as difficult and painful as breaking up with someone can ultimately be an act of loving. That's exactly what it was with Peter, and that act of loving has given me a friend whom I continue to love and support in my life.

This raises an even more interesting question about how we study meditation. If you have noticed, these kinds of transformations of self-worth or self-love were not mentioned once in the neurobiological studies of meditation. Those studies focused on brain waves or anatomical changes associated with meditation. From a neuroscience perspective, those are fine measures to start with, but my own experience with meditation suggests that there are much deeper and much more complex issues that one can examine with respect to the brain and meditation. All the changes that I experienced were examples of brain plasticity and seem to have some neurobiological basis. What is the network associated with enhanced self-love and acceptance? What happens when we shift our personal judgment system from an external-based model to an internal-based one? Also, I used a mixed bag of different kinds of meditation from Wayne Dyers's AH Meditation to Deepak Chopra's meditation challenges to loving kindness meditations. Which kind of meditation was doing what? By the way, my own mixed bag of meditation styles reflects the styles that have been used in published studies of meditation, making it difficult to compare across different practices. All this is to say that there is both an enormous potential for our understanding of the effects of meditation on brain function and an enormously challenging job ahead.

CLOSURE

You might be asking yourself about what happens when someone uses meditation to start living in the present and gains more love and appreciation for themselves. The answer is it heightens that person's awareness. And sometimes that leads to a beautiful example of closure.

This story starts with my desire to learn more about feng shui. Because I was trying to clear out all the stale old energy from the previous relationships in my life to make way for a wonderful new relationship, I thought this was the perfect time to give the process a try.

I had asked Inessa Freylekhman, a friend and feng shui expert, to come to my apartment. She arrived with burning sage, mantras to chant, and a set of beautiful gold bells from Bali that she used in the ceremony. She explained what each area of the house represents and identified where certain things needed to be moved or adjusted to make the energy flow better. After she evaluated the basic energy flow in my place (deemed good in that it could be improved with a few little fixes), we started walking through my apartment and she suggested that we look together at all the things that I had on my shelves and tables and ask if they represented old relationships and might need to be removed or represented good positive memories. It turned out that I had saved many mementos from those relationships in my life that had not worked out. Freylekhman suggested that I might want to find a new home for those items and refresh the energy in the rooms. I was stunned at the number of little knickknacks and reminders I had lining my walls and shelves, and I started feeling lighter as soon as we started identifying the things I could remove.

Then we got to my closet. It's a big walk-in closet, and when I opened it up the first thing we saw was a big cello in its case that took up a huge amount of my precious closet space. It was the same cello that François had given me all those years ago in France. Freylekhman asked what it was, and I explained. She asked if I played it, and I said no. Then she asked if I wanted to keep it.

I suddenly and totally unexpectedly started to cry.

She gently asked me what was wrong.

I told her that I felt so guilty for accepting this beautiful cello all those years ago, and not playing it. I told her I felt I didn't deserve it, and it made me feel like I was a terrible and totally unworthy recipient of this gorgeous gift. My guilt was compounded by my memory of the horrible way I broke up with François all those years ago over the phone. I have basically been dragging this guilt-riddled cello all around the country with me, not having time to play it and not having the heart to give it away. I hadn't even thought about it for years, yet there is was, making me burst into tears in the middle of my feng shui session.

I decided there and then that I wanted to give it away so it could do some good.

Freylekhman thought that was a great idea, and we finished our tour of my apartment and she finished all the blessings and mantras around my feng shui session. I would say it was a wonderful success.

I said good-bye to all the items we had identified and even a few more I found on my own and gave them all a new home. My place did feel lighter, airier, with a brand-new energy. I loved it.

Then I got to work on the cello. I learned that a colleague's daughter was looking for a new cello, and I enthusiastically suggested she could use mine. I loved the idea that it would be used. They got the cello evaluated by my friend's daughter's cello teacher, and that's when I discovered it had a big crack in it in a very bad place. My friend's daughter could not use it.

Then I learned that one of the students in our graduate program who is a cellist knew of a youth orchestra in need of usable cellos. That seemed like a lovely home for my instrument. They did not take broken cellos, but if I repaired it, they would be happy to take it. The student even gave me the name of his favorite cello-repair guy in the city.

I had everything I needed to implement my donation plan.

But then I did something very unlike me.

I did nothing.

I could not manage to find time to call the cello-repair guy no matter how hard I tried. Months passed until I realized what was wrong. I realized I still could not give my precious cello away—even to kids in need of instruments.

So the cello remained in the closet.

Then I went on a wine-tasting weekend with my friend Gina on the exotic North Fork of Long Island. She had heard that the wine tasting out there was surprisingly good. We both needed a break from the city, so off we went to a cute little bed-and-breakfast place for the weekend. She had big plans for a summer trip, and her friend had told her about a wonderful hotel in, of all places, Bordeaux, France. This place was on an estate with meals included, allowed access to the entire grounds, and was not very expensive. I mentioned my old boyfriend in Bordeaux with whom I had not spoken since college.

She said, wouldn't it be nice to stay in Bordeaux, and I could visit François and say hello. I was noncommittal because this whole conversation was making me realize something very important.

Yes, I did want to talk to François again, but not on a trip to Bordeaux. I realized that I needed to call and finally thank him properly not only for the cello but for that entire year in France. I knew what I was going to do.

When we got back from that Long Island weekend, I googled "piano tuners in Bordeaux" and didn't find much until I started looking at the images that came up and found a picture of him tuning a piano. He was looking a little older but it was definitely him. I found he worked for a recording studio and the next morning I woke up at 5:00 A.M. to call the studio to talk to him.

A man answered on the third ring, and in my rusty French, I asked if François was there.

He said, "Non."

I said, "Oh, doesn't he work there?"

He said, "Only when I need a piano tuner."

So he *did* know François!

I explained that I was an old friend from the United States and wanted to get in touch with him and asked if he would be able to give me François's cell phone number.

He said he had it, but it was at the shop, and he was at home at the moment. He asked me to call back in thirty minutes.

I thanked him and went back to bed for a nap.

About forty-five minutes later I called again, and there was no answer.

By this time my hopes were up that I would actually speak to

François so I was crushed when the man didn't answer the phone. But I told myself to be a little patient and I fixed myself a little breakfast and waited another thirty minutes or so.

My perseverance was rewarded because the next time I called he answered and had François's cell phone number ready for me. The guy from the studio was very kind and repeated the numbers at least four times to be sure I got them correct. I thanked him and hung up.

Without giving myself any time to think about the fact that I hadn't spoken to François for twenty-eight years and chicken out, I immediately called François's number and waited.

Someone picked up on the second ring.

I asked, "C'est François?"—Is this François?

He said, "Oui c'est moi."—Yes, it's me.

I said (in French), "Oh! This is an old friend from the United States; this is Wendy Suzuki."

He said (in French), "Hello!"

I said, "You don't seem that surprised to hear from me!"

He said that his friend from the recording studio had called him to tell him that an American woman was calling for him. He said I was the only American woman he knew, so he thought it was probably me.

We both laughed, and we had a nice time catching up with our families and our lives. His family is all well and he is now married with two daughters. I gave him a brief update about my family and my life.

He asked after my cello. I happily (and with a sense of relief) told him it was doing just fine.

That was the moment I took to tell him the real reason I called. I took a breath and said that because of the cello I realized that I had never properly thanked him for such an important experience in my life. I told him how special that whole year with him was to me. I told him (with a big lump in my throat) that I was calling him to finally say, "Thank you."

He was quiet for a moment.

All he said was, "Merci, Wendy."

He said the breakup had been very difficult for him and that year together had meant a lot to him as well. He said he was very happy to be in contact again after such a long time. We promised to be in touch via e-mail, wished each other the best, and hung up.

It was the ultimate example of closure in my life.

That conversion with François completely eliminated a huge piece of twenty-eight-year-old relationship baggage from my life. I had to be able to thank and acknowledge François for everything that year brought to me and all the wonderful new things it opened up to me to be able to fully appreciate anyone new coming into my life. That conversation together with my newly fung shui-ed apartment filled my home with a palpable new light and energy. It has also resulted in my repairing my beautiful cello—it's taking up a brand-new place of honor in my living room. Cello lessons are coming soon, but in the meantime I'm relearning how to tune it and play my scales, and it makes me smile every time I look at it.

Yes, I am truly ready to enjoy whatever comes next.

TAKE-AWAYS: THE EFFECTS OF MEDITATION ON BRAIN FUNCTION

- A major difference in the brain of expert meditators relative to novice meditators is a much higher level of gamma wave oscillations. Gamma wave oscillations have been linked to certain measures of consciousness.
- Expert meditators also have more efficient processing in brain areas important for attention.
- Eight weeks of meditation practice in novices has also been shown to be associated with strong EEG signals in the anterior part of the brain.

- Long-term meditation has been reported to increase the size of various brain areas.
- Meditation improves mood in subjects with social anxiety disorder (SAD).
- Meditation enhances attention during emotional situations in subjects with SAD.

FINAL NOTE

I n this book, I've shared with you the most meaningful insights
from my ongoing experiment with my own brain plasticity. One
thing I know for sure is that brain plasticity endows us with an
enormous capacity to change into the very best version of ourselves
that we can be. I've learned, largely through the process of writing
this book, that deep down inside I am the same little girl that loved
Broadway and math equally well. I've been through many different
phases in my life, from secret Broadway diva to painfully shy but
hardworking high school student to world-traveling college student
to total workaholic adult with no outside activities. All of those
phases, every one of them, were all me—just different versions of
me. I used to hate the shy high school nerd and the chubby workaholic assistant professor about to get tenure. Now I own them all.
Today, I am a balanced, joyful, intentional, and optimistic woman.
But just because I own them all doesn't mean that I don't want to
continue to change and grow. In particular, I know now that the
last big shift in my life from completely unbalanced neuroscientist
to happy and balanced woman happened because of (1) my clear

desire to improve myself; (2) hard work and perseverance, including a large dose of exercise; and (3) my own brain plasticity.

The most exciting news is that anyone can follow this program to transform his or her life. It starts with the innate desire to change. The other key ingredient you need in this mix is a little neuroscience know-how to point you in the right direction. With this book, I have tried to give you the tools that you need to start to improve your memory, attention, mood, and overall zest for life with aerobic and intentional exercise that will serve to kick-start your brain plasticity. What you do with that brain plasticity, including exactly how you apply it to different parts of your life, will be your own personal and unique creation.

Today, I am the best version of me that I have ever been. Exercise started me on the path that has allowed me to align my mind, body, and spirit in a way that I have never been able to do before. And I see the effects of those changes every day as I continue to explore how best I can serve in the world. And I'm not done changing and transforming—not by a long shot. I know I will continue to explore, grow, and change as I continue to express my own unique form of creativity for the rest of my life.

My wish is that my readers will use this book to inspire an exploration of their own remarkable brain plasticity and in doing so become the very best versions of themselves that they can be.

Wishing you all a healthy brain and a very happy life,
Wendy Suzuki
January 28, 2015

ACKNOWLEDGMENTS

I thank my coauthor Billie Fitzpatrick for her generosity, guidance, support, and superb writing skills. I could not have written this book without her amazing ability to find the heart of the story.

To Yfat Reiss Gendell, the very best book agent I could have ever asked for. I am so grateful for your knowledge, skill, and savvy regarding all things publishing. It's been a privilege to work with you.

I thank Carrie Thornton for her superb editorial skills. Every suggestion you made helped bring the book into better focus.

I thank and acknowledge my wonderful network of friends, my students, and my teachers who have supported me and inspired me throughout this process.

And I thank spirit for showing me my path of service.

REFERENCES

1: How a Geeky Girl Fell in Love with the Brain: The Science of Neuroplasticity and Enrichment

Anderson, A. K., and Phelps, E. A. "Lesions of the Human Amygdala Impair Enhanced Perception of Emotionally Salient Events." *Nature* 411 (2001): 305–309.

Bennett, E. L., Diamond, M. C., Krech, D., and Rosenzweig, M. R. "Chemical and Anatomical Plasticity of Brain." *Science* 146 (1964): 610–619.

Bennett, E. L., Rosenzweig, M. R., and Diamond, M. C. "Rat Brain: Effects of Environmental Enrichment on Wet and Dry Weights." *Science* 163 (1969): 825–826.

Blood, A. J., and Zatorre, R. J. "Intensely Pleasurable Responses to Music Correlate with Activity in Brain Regions Implicated in Reward and Emotion." *Proceedings of the National Academy of Sciences U.S.A.* 98 (2001): 11818–11823.

Cao, F., Tao, R., Liu, L., Perfetti, C. A., and Booth, J. R. "High Proficiency in a Second Language Is Characterized by Greater Involvement of the First Language Network: Evidence from Chinese Learners of English." *Journal of Cognitive Neuroscience* 25 (2013): 1649–1663.

Diamond, M. C., Krech, D., and Rosenzweig, M. R. "The Effects of an Enriched Environment on the Histology of the Rat Cerebral Cortex." *Journal of Comparative Neurology* 123 (1964): 111–120.

Diamond, M. C., Law, F., Rhodes, H., Lindner, B., Rosenzweig, M. R., Krech, D., and Bennett, E. L. "Increases in Cortical Depth and Glia Numbers in Rats Subjected to Enriched Environment." *Journal of Comparative Neurology* 128 (1966): 117–126.

Diamond, M. C., Rosenzweig, M. R., Bennett, E. L., Lindner, B., and Lyon, L. "Effects of Environmental Enrichment and Impoverishment on Rat Cerebral Cortex." *Journal of Neurobiology* 3 (1972): 47–64.

Globus, A., Rosenzweig, M. R., Bennett, E. L., and Diamond, M. C. "Effects of Differential Experience on Dendritic Spine Counts in Rat Cerebral Cortex." *Journal of Comparative Physiological Psychology* 82 (1973): 175–181.

Hamann, S. "Cognitive and Neural Mechanisms of Emotional Memory." *Trends in Cognitive Science* 5 (2001): 394–400.

Harrison, L., and Loui, P. "Thrills, Chills, Frissons, and Skin Orgasms: Toward an Integrative Model of Transcendent Psychophysiological Experiences in Music." *Frontiers in Psychology* 5 (2014): 790.

Hu, H., Real, E., Takamiya, K., Kang, M. G., Ledoux, J., Huganir, R. L., and Malinow, R. "Emotion Enhances Learning via Norepinephrine Regulation of AMPA-Receptor Trafficking." *Cell* 131 (2007): 160–173.

Klein, D., Mok, K., Chen, J. K., and Watkins, K. E. "Age of Language Learning Shapes Brain Structure: A Cortical Thickness Study of Bilingual and Monolingual Individuals." *Brain and Language* 131 (2014): 20–24.

Koelsch, S. "Towards a Neural Basis of Music-Evoked Emotions." *Trends in Cognitive Science* 14 (2010): 131–137.

Krech, D., Rosenzweig, M. R., and Bennett, E. L. "Effects of Environmental Complexity and Training on Brain Chemistry." *Journal of Comparative Physiological Psychology* 53 (1960): 509–519.

Kuhl, P. K. "Brain Mechanisms in Early Language Acquisition." *Neuron* 67 (2010): 713–727.

Kuhl, P. K. "Early Language Acquisition: Cracking the Speech Code." *Nature Reviews Neuroscience* 5 (2004): 831–843.

Kuhl, P. K., Stevens, E., Hayashi, A., Deguchi, T., Kiritani, S., and Iverson, P. "Infants Show a Facilitation Effect for Native Language Phonetic Perception between 6 and 12 Months." *Developmental Science* 9 (2006): F13–F21.

LaBar, K. S., and Cabeza, R. "Cognitive Neuroscience of Emotional Memory." *Nature Reviews Neuroscience* 7 (2006): 54–64.

Maguire, E. A., Gadian, D. G., Johnsrude, I. S., Good, C. D., Ashburner, J., Frackowiak, R. S., and Firth, C. D. "Navigation-Related Structural

Changes in the Hippocampus of Taxi Drivers." *Proceedings of the National Academy of Sciences U.S.A.* 98 (2000): 4398–4403.

Maguire, E. A., Spiers, H. J., Good, C. D., Hartley, T., Frackowiak, R. S., and Burgess, N. "Navigation Expertise and the Human Hippocampus: A Structural Brain Imaging Analysis." *Hippocampus* 13 (2003): 250–259.

Maguire, E. A., Woollett, K., and Spiers, H. J. "London Taxi Drivers and Bus Drivers: A Structural MRI and Neuropsychological Analysis." *Hippocampus* 16 (2006): 1091–1101.

Phelps, E. A. "Emotion and Cognition: Insights from Studies of the Human Amygdala." *Annual Review of Psychology* 57 (2006): 27–53.

Phelps, E. A. "Human Emotion and Memory: Interactions of the Amygdala and Hippocampal Complex." *Current Opinion in Neurobiology* 14 (2004): 198–202.

Phelps, E. A., and Anderson, A. K. "Emotional Memory: What Does the Amygdala Do?" *Current Biology* 7 (1997): R311–R314.

Rochefort, C., Gheusi, G., Vincent, J. D., and Lledo, P. M. "Enriched Odor Exposure Increases the Number of Newborn Neurons in the Adult Olfactory Bulb and Improves Odor Memory." *Journal of Neuroscience* 22 (2002): 2679–2689.

Rochefort, C., and Lledo, P. M. "Short-term Survival of Newborn Neurons in the Adult Olfactory Bulb after Exposure to a Complex Odor Environment." *European Journal of Neuroscience* 22 (2005): 2863–2870.

Rosenzweig, M. R., Krech, D., Bennett, E. L., and Diamond, M. C. "Effects of Environmental Complexity and Training on Brain Chemistry and Anatomy: A Replication and Extension." *Journal of Comparative Physiological Psychology* 55 (1962): 429–437.

Rosselli-Austin, L., and Williams, J. "Enriched Neonatal Odor Exposure Leads to Increased Numbers of Olfactory Bulb Mitral and Granule Cells." *Brain Research. Developmental Brain Research* 51 (1990): 135–137.

Salimpoor, V. N., Benovoy, M., Larcher, K., Dagher, A., and Zatorre, R. J. "Anatomically Distinct Dopamine Release during Anticipation and Experience of Peak Emotion to Music." *Nature Neuroscience* 14 (2011): 257–262.

Sharot, T., Delgado, M. R., and Phelps, E. A. "How Emotion Enhances the Feeling of Remembering." *Nature Neuroscience* 7 (2004): 1376–1380.

Woollett, K., and Maguire, E. A. "Acquiring 'the Knowledge' of London's Layout Drives Structural Brain Changes." *Current Biology* 21 (2011): 2109–2114.

Zatorre, R. J., and Salimpoor, V. N. "From Perception to Pleasure: Music and Its Neural Substrates." *Proceedings of the National Academy of Sciences U.S.A.* 110 Suppl. 2 (2013): 10430–10437.

2: Solving the Mysteries of Memory: How Memories Are Formed and Retained

Amaral, D. G., Insausti, R., Zola-Morgan, S., Squire, L. R., and Suzuki, W. A. "The Perirhinal and Parahippocampal Cortices in Memory Function." In *Proceedings of Vision, Memory and the Temporal Lobe,* edited by Mortimer Mishkin and Eiichi Iwai, 149–161. New York: Elsevier Science Ltd., 1990.

Corkin, S. "Acquisition of Motor Skill after Bilateral Medial Temporal-Lobe Excision." *Neuropsychologia* 6 (1968): 255–265.

Corkin, S. *Permanent Present Tense.* New York: Basic Books, 2013.

Corkin, S., Amaral, D. G., Gilberto Gonzalez, R., Johnson, K. A., and Hyman, B. T. "H.M.'S Medial Temporal Lobe Lesion: Findings from Magnetic Resonance Imaging." *Journal of Neuroscience* 17 (1997): 3964–3979.

Insausti, R., Amaral, D. G., and Cowan, W. M. "The Entorhinal Cortex of the Monkey: II. Cortical Afferents." *Journal of Comparative Neurology* 264 (1987): 356–395.

Lashley, K. S. *Brain Mechanisms and Intelligence: A Quantitative Study of Injuries to the Brain.* Chicago: Chicago University Press, 1929.

Lashley, K. S. "Mass Action in Cerebral Function." *Science* 73 (1931): 245–254.

Milner, B. "Disorders of Learning and Memory after Temporal Lobe Lesions in Man." *Neuropsychologia* 19 (1972): 421–446.

Milner, B. "The Medial Temporal-Lobe Amnesic Syndrome." *Psychiatric Clinics of North America* 28 (2005): 599–611.

Milner, B. "The Memory Defect in Bilateral Hippocampal Lesions." *Psychiatric Research Reports* 11 (1959): 43–58.

Milner, B. "Wilder Penfield: His Legacy to Neurology. Memory Mechanisms." *Canadian Medical Association Journal* 116 (1977): 1374–1376.

Milner, B., and Penfield, W. "The Effect of Hippocampal Lesions on Recent Memory." *Transactions of the American Neurological Association* 80 (1955): 42–48.

Milner, B., Squire, L. R., and Kandel, E. R. "Cognitive Neuroscience and the Study of Memory." *Neuron* 20 (1998): 445–468.

Mishkin, M. "Memory in Monkeys Severely Impaired by Combined but Not by Separate Removal of Amygdala and Hippocampus." *Nature* 273 (1978): 297–298.

Penfield, W., and Mathieson, G. "Memory: Autopsy Findings and Comments on the Role of Hippocampus in Experiential Recall." *Archives of Neurology* 31 (1974): 145–154.

Penfield. W., and Milner, B. "Memory Deficit Produced by Bilateral Lesions in the Hippocampal Zone." *A.M.A. Archives of Neurology and Psychiatry* 79 (1958): 475–497.

Scoville, W. B., and Milner, B. "Loss of Recent Memory after Bilateral Hippocampal Lesions." *Journal of Neurology, Neurosurgery, and Psychology* 20 (1957): 11–21.

Squire, L. R. "Declarative and Nondeclarative Memory: Multiple Brain Systems Supporting Learning and Memory." *Journal of Cognitive Neuroscience* 4 (1992): 232–243.

Squire, L. R. "Memory Systems of the Brain: A Brief History and Current Perspective." *Neurobiology of Learning and Memory* 82 (2004): 171–177.

Squire, L. R., Clark, R. E., and Bailey, P. J. "Medial Temporal Lobe Function and Memory." In *The Cognitive Neurosciences III*, edited by M. Gazzaniga, 691–708. Cambridge: MIT Press, 2004.

Squire, L. R., and Knowlton, B. J. "Memory, Hippocampus, and Brain Systems." In *The Cognitive Neurosciences*, edited by M. Gazzaniga, 825–837. Cambridge: MIT Press, 1994.

Squire, L. R., and Zola, S. M. "Episodic Memory, Semantic Memory and Amnesia." *Hippocampus* 8 (1998): 205–211.

Suzuki, W. A. "Neuroanatomy of the Monkey Entorhinal, Perirhinal and Parahippocampal Cortices: Organization of Cortical Inputs and Interconnections with Amygdala and Striatum." *Neurosciences* 8 (1996): 3–12.

Suzuki, W. A., and Amaral, D. G. "Cortical Inputs to the CA1 Field of the Monkey Hippocampus Originate from the Perirhinal and Parahippocampal Cortex but Not from Area TE." *Neuroscience Letters* 115 (1990): 43–48.

Suzuki, W. A., and Amaral, D. G. "Functional Neuroanatomy of the Medial Temporal Lobe Memory System." *Cortex* 40 (2004): 220–222.

Suzuki, W. A., and Amaral, D. G. "Perirhinal and Parahippocampal Cortices of the Macaque Monkey: Cortical Afferents." *Journal of Comparative Neurology* 350 (1994): 497–533.

Suzuki, W. A., and Amaral, D. G. "The Perirhinal and Parahippocampal Cortices of the Macaque Monkey: Cytoarchitectonic and Chemoarchitectonic Organization." *Journal of Comparative Neurology* 463 (2003): 67–91.

Suzuki, W. A., and Amaral, D. G. "Topographic Organization of the

Reciprocal Connections between Monkey Entorhinal Cortex and the Perirhinal and Parahippocampal Cortices." *Journal of Neuroscience* 14 (1994): 1856–1877.

Suzuki, W. A., and Amaral, D. G. "Where Are the Perirhinal and Parahippocampal Cortices? A Historical Overview of the Nomenclature and Boundaries Applied to the Primate Medial Temporal Lobe." *Neuroscience* 120 (2003): 893–906.

Suzuki, W. A., and Brown, E. N. "Behavioral and Neurophysiological Analyses of Dynamic Learning Processes." *Behavioral and Cognitive Neuroscience Reviews* 4 (2005): 67–95.

Suzuki, W. A., Miller, E. K., and Desimone, R. "Object and Place Memory in the Macaque Entorhinal Cortex." *Journal of Neuroscience* 78 (1997): 1062–1081.

Suzuki, W. A., Zola-Morgan, S., Squire, L. R., and Amaral, D. G. "Lesions of the Perirhinal and Parahippocampal Cortices in the Monkey Produce Long-Lasting Memory Impairment in the Visual and Tactual Modalities." *Journal of Neuroscience* 13 (1993): 2430–2451.

Wirth, S., Yanike, M., Frank, L. M., Smith, A. C., Brown, E. N., and Suzuki, W. A. "Single Neurons in the Monkey Hippocampus and Learning of New Associations." *Science* 300 (2003): 1578–1581.

Zola-Morgan, S., Squire, L. R., and Amaral, D. G. "Lesions of the Amygdala That Spare Adjacent Cortical Regions Do Not Impair Memory or Exacerbate the Impairment Following Lesions of the Hippocampal Formation." *Journal of Neuroscience* 9 (1989): 1922–1936.

Zola-Morgan, S., Squire, L. R., Amaral, D. G., and Suzuki, W. A. "Lesions of Perirhinal and Parahippocampal Cortex That Spare the Amygdala and Hippocampal Formation Produce Severe Memory Impairment." *Journal of Neuroscience* 9 (1989): 4355–4370.

Zola-Morgan, S., Squire, L. R., and Mishkin, M. "The Neuroanatomy of Amnesia: Amygdala-Hippocampus Versus Temporal Stem." *Science* 218 (1982): 1337–1339.

3: The Mystery of Memory Hits Home: Memories Mean More Than Neurons

Anderson, A. K., and Phelps, E. A. "Lesions of the Human Amygdala Impair Enhanced Perception of Emotionally Salient Events." *Nature* 411 (2001): 305–309.

Hamann, S. "Cognitive and Neural Mechanisms of Emotional Memory." *Trends in Cognitive Sciences* 5 (2001): 394–400.

Hu, H., Real, E., Takamiya, K., Kang, M. G., Ledoux, J., Huganir, R. L., and Malinow, R. "Emotion Enhances Learning Via Norepinephrine Regulation of AMPA-Receptor Trafficking." *Cell* 131 (2007): 160–173.

LaBar, K. S., and Cabeza, R. "Cognitive Neuroscience of Emotional Memory." *Nature Reviews Neuroscience.* 7 (2006): 54–64.

Phelps, E. A. "Emotion and Cognition: Insights from Studies of the Human Amygdala." *Annual Review of Psychology* 57 (2006): 27–53.

Phelps, E. A. "Human Emotion and Memory: Interactions of the Amygdala and Hippocampal Complex." *Current Opinion in Neurobiology* 14 (2004): 198–202.

Phelps, E. A., and Anderson, A. K. "Emotional Memory: What Does the Amygdala Do?" *Current Biology* 7 (1997): R311–R314.

Sharot, T., Delgado, M. R., and Phelps, E. A. "How Emotion Enhances the Feeling of Remembering." *Nature Neuroscience* 7 (2004): 1376–1380.

4: Chunky No More: Reconnecting My Brain with My Body and Spirit

Arnone, D., McKie, S., Elliott, R., Juhasz, G., Thomas, E. J., Downey, D., Williams, S., Deakin, J. F., and Anderson, I. M. "State-Dependent Changes in Hippocampal Grey Matter in Depression." *Molecular Psychiatry* 18 (2013): 1265–1272.

Arnone, D., McKie, S., Elliott, R., Thomas, E. J., Downey, D., Juhasz, G., Williams, S. R., Deakin, J. F., and Anderson, I. M. "Increased Amygdala Responses to Sad but Not Fearful Faces in Major Depression: Relation to Mood State and Pharmacological Treatment." *American Journal of Psychiatry* 169 (2012): 841–850.

Boecker, H., Sprenger, T., Spilker, M. E., Henriksen, G., Koppenhoefer, M., Wagner, K. J., Valet, M., Berthele, A., and Tolle, T. R. "The Runner's High: Opioidergic Mechanisms in the Human Brain." *Cerebral Cortex* 18 (2008): 2523–2531.

Carney, D. R., Cuddy, A. J., and Yap, A. J. "Power Posing: Brief Nonverbal Displays Affect Neuroendocrine Levels and Risk Tolerance." *Psychological Science* 21 (2010): 1363–1368.

Chennaoui, M., Grimaldi, B., Fillion, M. P., Bonnin, A., Drogou, C., Fillion, G., and Guezennec, C. Y. "Effects of Physical Training on Functional Activity of 5-HT1B Receptors in Rat Central Nervous System: Role of 5-HT-Moduline." *Naunyn-Schmiedeberg's Archives of Pharmacology* 361 (2000): 600–604.

Cohen, G. L., and Sherman, D. K. "The Psychology of Change:

Self-Affirmation and Social Psychological Intervention." *Annual Review of Psychology* 65 (2014): 333–371.

Cotman, C. W., and Engesser-Cesar, C. "Exercise Enhances and Protects Brain Function." *Exercise and Sport Science Reviews* 30 (2002): 75–79.

de Castro, J. M., and Duncan, G. "Operantly Conditioned Running: Effects on Brain Catecholamine Concentrations and Receptor Densities in the Rat." *Pharmacology, Biochemistry, and Behavior* 23 (1985): 495–500.

Dunn, A. L., Reigle, T. G., Youngstedt, S. D., Armstrong, R. B., and Dishman, R. K. "Brain Norepinephrine and Metabolites after Treadmill Training and Wheel Running in Rats." *Medicine and Science in Sports and Exercise* 28 (1996): 204–209.

Gauvin, L., Rejeski, W. J., and Norris, J. L. "A Naturalistic Study of the Impact of Acute Physical Activity on Feeling States and Affect in Women." *Health Psychology* 15 (1996): 391–397.

Koenigs, M., and Grafman, J. "Posttraumatic Stress Disorder: The Role of Medial Prefrontal Cortex and Amygdala." *Neuroscientist* 15 (2009): 540–548.

Lin, T. W., and Kuo, Y. M. "Exercise Benefits Brain Function: The Monoamine Connection." *Brain Sciences* 3 (2013): 39–53.

Lindsay, E. K., and Creswell, J. D. "Helping the Self Help Others: Self-Affirmation Increases Self-Compassion and Pro-Social Behaviors." *Frontiers in Psychology* 5 (2014): 421.

Lorenzetti, V., Allen, N. B., Fornito, A., and Yucel, M. "Structural Brain Abnormalities in Major Depressive Disorder: A Selective Review of Recent MRI Studies." *Journal of Affective Disorders* 117 (2009): 1–17.

Lorenzetti, V., Allen, N. B., Whittle, S., and Yucel, M. "Amygdala Volumes in a Sample of Current Depressed and Remitted Depressed Patients and Healthy Controls." *Journal of Affective Disorders* 120 (2010): 112–119.

Masi, G., and Brovedani, P. "The Hippocampus, Neurotrophic Factors and Depression: Possible Implications for the Pharmacotherapy of Depression." *CNS Drugs* 25 (2011): 913–931.

Rejeski, W. J., Gauvin, L., Hobson, M. L., and Norris, J. L. "Effects of Baseline Responses, In-Task Feelings, and Duration of Activity on Exercise-Induced Feeling States in Women." *Health Psychology* 14 (1995): 350–359.

Salmon, P. "Effects of Physical Exercise on Anxiety, Depression, and Sensitivity to Stress: A Unifying Theory." *Clinic Psychological Review* 21 (2001): 33–61.

Steptoe, A., Kimbell, J., and Basford, P. "Exercise and the Experience and Appraisal of Daily Stressors: A Naturalistic Study." *Journal of Behavioral Medicine* 21 (1998): 363–374.

Tuson, K. M., Sinyor, D., and Pelletier, L. G. "Acute Exercise and Positive Affect: An Investigation of Psychological Processes Leading to Affective Change." *International Journal of Sport Psychology* 26 (1995): 138–159.

Villanueva, R. "Neurobiology of Major Depressive Disorder." *Neural Plasticity* 2013 (2013): 873278.

Yeung, R. R. "The Acute Effects of Exercise on Mood State." *Journal of Psychosomatic Research* 40 (1996): 123–141.

Young, S. N. "How to Increase Serotonin in the Human Brain without Drugs." *Journal of Psychiatry and Neuroscience* 32 (2007): 394–399.

Young, S. N., and Leyton, M. "The Role of Serotonin in Human Mood and Social Interaction. Insight from Altered Tryptophan Levels." *Pharmacology, Biochemistry, and Behavior* 71 (2002): 857–865.

5: The Birth of an Idea: How Does Exercise *Really* Affect the Brain?

Altman, J. "Are New Neurons Formed in the Brains of Adult Mammals?" *Science* 135 (1962): 1127–1128.

Altman, J., and Das, G. D. "Autoradiographic and Histological Evidence of Postnatal Hippocampal Neurogenesis in Rats." *Journal of Comparative Neurology* 124 (1965): 319–335.

Boecker, H., Sprenger, T., Spilker, M. E., Henriksen, G., Koppenhoefer, M., Wagner, K. J., Valet, M., Berthele, A., and Tolle, T. R. "The Runner's High: Opioidergic Mechanisms in the Human Brain." *Cerebral Cortex* 18 (2008): 2523–2531.

Brickman, A. M., Stern, Y., and Small, S. A. "Hippocampal Subregions Differentially Associate with Standardized Memory Tests." *Hippocampus* 21 (2011): 923–928.

Brown, J., Cooper-Kuhn, C. M., Kempermann, G., Van, P. H., Winkler, J., Gage, F. H., and Kuhn, H. G. "Enriched Environment and Physical Activity Stimulate Hippocampal but Not Olfactory Bulb Neurogenesis." *European Journal of Neuroscience* 17 (2003): 2042–2046.

Cassilhas, R. C., Lee, K. S., Fernandes, J., Oliveira, M. G., Tufik, S., Meeusen, R., and de Mello, M. T. "Spatial Memory Is Improved by Aerobic and Resistance Exercise through Divergent Molecular Mechanisms." *Neuroscience* 202 (2012): 309–317.

Chennaoui, M., Grimaldi, B., Fillion, M. P., Bonnin, A., Drogou, C., Fillion, G., and Guezennec, C. Y. "Effects of Physical Training on

Functional Activity of 5-HT1B Receptors in Rat Central Nervous System: Role of 5-HT-Moduline." *Naunyn-Schmiedeberg's Archives of Pharmacology* 361 (2000): 600–604.

Colcombe, S., and Kramer, A. F. "Fitness Effects on the Cognitive Function of Older Adults: A Meta-Analytic Study." *Psychological Science* 14 (2003): 125–130.

Creer, D. J., Romberg, C., Saksida, L. M., Van, P. H., and Bussey, T. J. "Running Enhances Spatial Pattern Separation in Mice." *Proceedings of the National Academy of Sciences U.S.A.* 107 (2010): 2367–2372.

de Castro, J. M., and Duncan, G. "Operantly Conditioned Running: Effects on Brain Catecholamine Concentrations and Receptor Densities in the Rat." *Pharmacology, Biochemistry, and Behavior* 23 (1985): 495–500.

Dunn, A. L., Reigle, T. G., Youngstedt, S. D., Armstrong, R. B., and Dishman, R. K. "Brain Norepinephrine and Metabolites after Treadmill Training and Wheel Running in Rats." *Medicine and Science in Sports and Exercise* 28 (1996): 204–209.

Erickson, K. I., Voss, M. W., Prakash, R. S., Basak, C., Szabo, A., Chaddock, L., Kim, J. S., Heo, S., Alves, H., White, S. M., Wojcicki, T. R., Mailey, E., Vieira, V. J., Martin, S. A., Pence, B. D., Woods, J. A., McAuley, E., and Kramer, A. F. "Exercise Training Increases Size of Hippocampus and Improves Memory." *Proceedings of the National Academy of Sciences U.S.A.* 108 (2011): 3017–3022.

Eriksson, P. S., Perfilieva, E., Bjork-Eriksson, T., Alborn, A. M., Nordborg, C., Peterson, D. A., and Gage, F. H. "Neurogenesis in the Adult Human Hippocampus." *Nature Medicine* 4 (1998): 1313–1317.

Frick, K. M., and Fernandez, S. M. "Enrichment Enhances Spatial Memory and Increases Synaptophysin Levels in Aged Female Mice." *Neurobiology of Aging* 24 (2003): 615–626.

Gross, C. G. "Neurogenesis in the Adult Brain: Death of a Dogma." *Nature Reviews Neuroscience* 1 (2000): 67–73.

Hillman, C. H., Erickson, K. I., and Kramer, A. F. "Be Smart, Exercise Your Heart: Exercise Effects on Brain and Cognition." *Nature Reviews Neuroscience* 9 (2008): 58–65.

Hopkins, M. E., and Bucci, D. J. "BDNF Expression in Perirhinal Cortex Is Associated with Exercise-Induced Improvement in Object Recognition Memory." *Neurobiology of Learning and Memory* 94 (2010): 278–284.

Ickes, B. R., Pham, T. M., Sanders, L. A., Albeck, D. S., Mohammed, A. H., and Granholm, A. C. "Long-Term Environmental Enrichment Leads to Regional Increases in Neurotrophin Levels in Rat Brain." *Experimental Neurology* 164 (2000): 45–52.

Jung, C. K., and Herms, J. "Structural Dynamics of Dendritic Spines Are Influenced by an Environmental Enrichment: An In Vivo Imaging Study." *Cerebral Cortex* 24 (2014): 377–384.

Kempermann, G., Kuhn, H. G., and Gage, F. H. "More Hippocampal Neurons in Adult Mice Living in an Enriched Environment." *Nature* 386 (1997): 493–495.

Kleim, J. A., Cooper, N. R., and VandenBerg, P. M. "Exercise Induces Angiogenesis but Does Not Alter Movement Representations within Rat Motor Cortex." *Brain Research* 934 (2002): 1–6.

Kobilo, T., Liu, Q. R., Gandhi, K., Mughal, M., Shaham, Y., and Van, P. H. "Running Is the Neurogenic and Neurotrophic Stimulus in Environmental Enrichment." *Learning and Memory* 18 (2011): 605–609.

Krech, D., Rosenzweig, M. R., and Bennett, E. L. "Effects of Environmental Complexity and Training on Brain Chemistry." *Journal of Comparative and Physiological Psychology* 53 (1960): 509–519.

Larson, E. B., Wang, L., Bowen, J. D., McCormick, W. C., Teri, L., Crane, P., and Kukull, W. "Exercise Is Associated with Reduced Risk for Incident Dementia among Persons 65 Years of Age and Older." *Annals of Internal Medicine* 144 (2006): 73–81.

Lin, T. W., and Kuo, Y. M. "Exercise Benefits Brain Function: The Monoamine Connection." *Brain Science* 3 (2013): 39–53.

Marlatt, M. W., Potter, M. C., Lucassen, P. J., and Van, P. H. "Running throughout Middle-Age Improves Memory Function, Hippocampal Neurogenesis, and BDNF Levels in Female C57BL/6J Mice." *Developmental Neurobiology* 72 (2012): 943–952.

Pereira, A. C., Huddleston, D. E., Brickman, A. M., Sosunov, A. A., Hen, R., McKhann, G. M., Sloan, R., Gage, F. H., Brown, T. R., and Small, S. A. "An In Vivo Correlate of Exercise-Induced Neurogenesis in the Adult Dentate Gyrus." *Proceedings of the National Academy of Sciences U.S.A.* 104 (2007): 5638–5643.

Rosenzweig, M. R., Krech, D., Bennett, E. L., and Diamond, M. C. "Effects of Environmental Complexity and Training on Brain Chemistry and Anatomy: A Replication and Extension." *Journal of Comparative Physiological Psychology* 55 (1962): 429–437.

Smith, P. J., Blumenthal, J. A., Hoffman, B. M., Cooper, H., Strauman, T. A., Welsh-Bohmer, K., Browndyke, J. N., and Sherwood, A. "Aerobic Exercise and Neurocognitive Performance: A Meta-Analytic Review of Randomized Controlled Trials." *Psychosomatic Medicine* 72 (2010): 239–252.

Stranahan, A. M., Khalil, D., and Gould, E. "Running Induces Widespread Structural Alterations in the Hippocampus and Entorhinal Cortex." *Hippocampus* 17 (2007): 1017–1022.

Van, P. H. "Neurogenesis and Exercise: Past and Future Directions." *Neuromolecular Medicine* 10 (2008): 128–140.

Van, P. H., Christie, B. R., Sejnowski, T. J., and Gage, F. H. "Running Enhances Neurogenesis, Learning, and Long-Term Potentiation in Mice." *Proceedings of the National Academy of Sciences U.S.A.* 96 (1999): 13427–13431.

Van, P. H., Kempermann, G., and Gage, F. H. "Neural Consequences of Environmental Enrichment." *Nature Reviews Neuroscience* 1 (2000): 191–198.

Van, P. H., Kempermann, G., and Gage, F. H. "Running Increases Cell Proliferation and Neurogenesis in the Adult Mouse Dentate Gyrus." *Nature Neuroscience* 2 (1999): 266–270.

Van, P. H., Shubert, T., Zhao, C., and Gage, F. H. "Exercise Enhances Learning and Hippocampal Neurogenesis in Aged Mice." *Journal of Neuroscience* 25 (2005): 8680–8685.

Voss, M. W., Vivar, C., Kramer, A. F., and Van, P. H. "Bridging Animal and Human Models of Exercise-Induced Brain Plasticity." *Trends in Cognitive Science* 17 (2013): 525–544.

Young, S. N. "How to Increase Serotonin in the Human Brain without Drugs." *Journal of Psychiatry and Neuroscience* 32 (2007): 394–399.

Young, S. N., and Leyton, M. "The Role of Serotonin in Human Mood and Social Interaction. Insight from Altered Tryptophan Levels." *Pharmacology, Biochemistry, and Behavior* 71 (2002): 857–865.

6: Spandex in the Classroom: Exercise Can Make You Smarter

Colcombe, S., and Kramer, A. F. "Fitness Effects on the Cognitive Function of Older Adults: A Meta-Analytic Study." *Psychological Science* 14 (2003): 125–130.

Creer, D. J., Romberg, C., Saksida, L. M., Van, P. H., and Bussey, T. J. "Running Enhances Spatial Pattern Separation in Mice." *Proceedings of the National Academy of Sciences U.S.A.* 107 (2010): 2367–2372.

Kinser, P. A., Goehler, L. E., and Taylor, A. G. "How Might Yoga Help Depression? A Neurobiological Perspective." *Explore* 8 (2012): 118–126.

Lee, Y. S., Ashman, T., Shang, A., and Suzuki, W. "Brief Report: Effects of Exercise and Self-Affirmation Intervention after Traumatic Brain Injury." *NeuroRehabilitation* 35 (2014): 57–65.

Rocha, K. K., Ribeiro, A. M., Rocha, K. C., Sousa, M. B., Albuquerque, F. S., Ribeiro, S., and Silva, R. H. "Improvement in Physiological and Psychological Parameters after 6 Months of Yoga Practice." *Consciousness and Cognition* 21 (2012): 843–850.

Thomley, B. S., Ray, S. H., Cha, S. S., and Bauer, B. A. "Effects of a Brief, Comprehensive, Yoga-Based Program on Quality of Life and Biometric Measures in an Employee Population: A Pilot Study." *Explore* 7 (2011): 27–29.

Voss, M. W., Nagamatsu, L. S., Liu-Ambrose, T., and Kramer, A. F. "Exercise, Brain, and Cognition across the Life Span." *Journal of Applied Physiology* 111 (2011): 1505–1513.

7: I Stress, You Stress, We All Stress! Challenging the Neurobiology of Stress Response

Adlard, P. A., and Cotman, C. W. "Voluntary Exercise Protects against Stress-Induced Decreases in Brain-Derived Neurotrophic Factor Protein Expression." *Neuroscience* 124 (2004): 985–992.

Ansell, E. B., Rando, K., Tuit, K., Guarnaccia, J., and Sinha, R. "Cumulative Adversity and Smaller Gray Matter Volume in Medial Prefrontal, Anterior Cingulate, and Insula Regions." *Biological Psychiatry* 72 (2012): 57–64.

Arnsten, A. F. "Stress Signalling Pathways That Impair Prefrontal Cortex Structure and Function." *Nature Reviews Neuroscience* 10 (2009): 410–422.

Baek, S. S., Jun, T. W., Kim, K. J., Shin, M. S., Kang, S. Y., and Kim, C. J. "Effects of Postnatal Treadmill Exercise on Apoptotic Neuronal Cell Death and Cell Proliferation of Maternal-Separated Rat Pups." *Brain and Development* 34 (2012): 45–56.

Bannerman, D. M., Rawlins, J. N., McHugh, S. B., Deacon, R. M., Yee, B. K., Bast, T., Zhang, W. N., Pothuizen, H. H., and Feldon, J. "Regional Dissociations within the Hippocampus—Memory and Anxiety." *Neuroscience and Biobehavioral Reviews* 28 (2004): 273–283.

Barbour, K. A., Edenfield, T. M., and Blumenthal, J. A. "Exercise as a Treatment for Depression and Other Psychiatric Disorders: A Review." *Journal of Cardiopulmonary Rehabilitation and Prevention* 27 (2007): 359–367.

Blumenthal, J. A., Smith, P. J., and Hoffman, B. M. "Is Exercise a Viable Treatment for Depression?" *ACSM's Health and Fitness Journal* 16 (2012): 14–21.

Bremner, J. D., Narayan, M., Staib, L. H., Southwick, S. M., McGlashan, T., and Charney, D. S. "Neural Correlates of Memories of Childhood Sexual

Abuse in Women with and without Posttraumatic Stress Disorder." *American Journal of Psychiatry* 156 (1999): 1787–1795.

Bremner, J. D., Staib, L. H., Kaloupek, D., Southwick, S. M., Soufer, R., and Charney, D. S. "Neural Correlates of Exposure to Traumatic Pictures and Sound in Vietnam Combat Veterans with and without Posttraumatic Stress Disorder: A Positron Emission Tomography Study." *Biological Psychiatry* 45 (1999): 806–816.

Davidson, R. J., and McEwen, B. S. "Social Influences on Neuroplasticity: Stress and Interventions to Promote Well-Being." *Nature Neuroscience* 15 (2012): 689–695.

Driessen, M., Beblo, T., Mertens, M., Piefke, M., Rullkoetter, N., Silva-Saavedra, A., Reddemann, L., Rau, H., Markowitsch, H. J., Wulff, H., Lange, W., and Woermann, F. G. "Posttraumatic Stress Disorder and fMRI Activation Patterns of Traumatic Memory in Patients with Borderline Personality Disorder." *Biological Psychiatry* 55 (2004): 603–611.

Hendler, T., Rotshtein, P., Yeshurun, Y., Weizmann, T., Kahn, I., Ben-Bashat, D., Malach, R., and Bleich, A. "Sensing the Invisible: Differential Sensitivity of Visual Cortex and Amygdala to Traumatic Context." *Neuroimage* 19 (2003): 587–600.

Herring, M. P., O'Connor, P. J., and Dishman, R. K. "The Effect of Exercise Training on Anxiety Symptoms Among Patients: A Systematic Review." *Archives of Internal Medicine* 170 (2010): 321–331.

Hoffman, B. M., Babyak, M. A., Craighead, W. E., Sherwood, A., Doraiswamy, P. M., Coons, M. J., and Blumenthal, J. A. "Exercise and Pharmacotherapy in Patients with Major Depression: One-Year Follow-Up of the SMILE Study." *Psychosomatic Medicine* 73 (2011): 127–133.

Kannangara, T. S., Webber, A., Gil-Mohapel, J., and Christie, B. R. "Stress Differentially Regulates the Effects of Voluntary Exercise on Cell Proliferation in the Dentate Gyrus of Mice." *Hippocampus* 19 (2009): 889–897.

Kempermann, G., and Kronenberg, G. "Depressed New Neurons—Adult Hippocampal Neurogenesis and a Cellular Plasticity Hypothesis of Major Depression." *Biological Psychiatry* 54 (2003): 499–503.

Kitayama, N., Vaccarino, V., Kutner, M., Weiss, P., and Bremner, J. D. "Magnetic Resonance Imaging (MRI) Measurement of Hippocampal Volume in Posttraumatic Stress Disorder: A Meta-Analysis." *Journal of Affective Disorders* 88 (2005): 79–86.

Koenigs, M., and Grafman, J. "Posttraumatic Stress Disorder: The Role of Medial Prefrontal Cortex and Amygdala." *Neuroscientist* 15 (2009): 540–548.

Levine, S. "Plasma-Free Corticosteroid Response to Electric Shock in Rats Stimulated in Infancy." *Science* 135 (1962): 795–796.

Lyons, D. M., Parker, K. J., and Schatzberg, A. F. "Animal Models of Early Life Stress: Implications for Understanding Resilience." *Developmental Psychobiology* 52 (2010): 616–624.

McEwen, B. S. "Physiology and Neurobiology of Stress and Adaptation: Central Role of the Brain." *Physiological Reviews* 87 (2007): 873–904.

McEwen, B. S., and Morrison, J. H. "The Brain on Stress: Vulnerability and Plasticity of the Prefrontal Cortex over the Life Course." *Neuron* 79 (2013): 16–29.

Ondicova, K., and Mravec, B. "Multilevel Interactions between the Sympathetic and Parasympathetic Nervous Systems: A Minireview." *Endocrine Regulations* 44 (2010): 69–75.

Parker, K. J., Buckmaster, C. L., Justus, K. R., Schatzberg, A. F., and Lyons, D. M. "Mild Early Life Stress Enhances Prefrontal-Dependent Response Inhibition in Monkeys." *Biological Psychiatry* 57 (2005): 848–855.

Parker, K. J., Buckmaster, C. L., Schatzberg, A. F., and Lyons, D. M. "Prospective Investigation of Stress Inoculation in Young Monkeys." *Archives of General Psychiatry* 61 (2004): 933–941.

Paton, J. F., Boscan, P., Pickering, A. E., and Nalivaiko, E. "The Yin and Yang of Cardiac Autonomic Control: Vago-Sympathetic Interactions Revisited." *Brain Research. Brain Research Reviews* 49 (2005): 555–565.

Perraton, L. G., Kumar, S., and Machotka, Z. "Exercise Parameters in the Treatment of Clinical Depression: A Systematic Review of Randomized Controlled Trials." *Journal of Evaluation in Clinical Practice* 16 (2010): 597–604.

Russo, S. J., Murrough, J. W., Han, M. H., Charney, D. S., and Nestler, E. J. "Neurobiology of Resilience." *Nature Neuroscience* 15 (2012): 1475–1484.

Sahay, A., Drew, M. R., and Hen, R. "Dentate Gyrus Neurogenesis and Depression." *Progress in Brain Research* 163 (2007): 697–722.

Sahay, A., and Hen, R. "Adult Hippocampal Neurogenesis in Depression." *Nature Neuroscience* 10 (2007): 1110–1115.

Sapolsky, R. M. "The Influence of Social Hierarchy on Primate Health." *Science* 308 (2005): 648–652.

Sapolsky, R. M. "Taming Stress." *Scientific American* 289, no. 3 (2003): 86–95.

Sapolsky, R. M. "Why Stress Is Bad for Your Brain." *Science* 273 (1996): 749–750.

Schoenfeld, T. J., and Gould, E. "Stress, Stress Hormones, and Adult Neurogenesis." *Experimental Neurology* 233 (2012): 12–21.

Shin, L. M., Orr, S. P., Carson, M. A., Rauch, S. L., Macklin, M. L., Lasko, N. B., Peters, P. M., Metzger, L. J., Dougherty, D. D., Cannistraro, P. A., Alpert, N. M., Fischman, A. J., and Pitman, R. K. "Regional Cerebral Blood Flow in the Amygdala and Medial Prefrontal Cortex during Traumatic Imagery in Male and Female Vietnam Veterans with PTSD." *Archives of General Psychiatry* 61 (2004): 168–176.

Uno, H., Tarara, R., Else, J. G., Suleman, M. A., and Sapolsky, R. M. "Hippocampal Damage Associated with Prolonged and Fatal Stress in Primates." *Journal of Neuroscience* 9 (1989): 1705–1711.

Woon, F. L., Sood, S., and Hedges, D. W. "Hippocampal Volume Deficits Associated with Exposure to Psychological Trauma and Posttraumatic Stress Disorder in Adults: A Meta-Analysis." *Progress in Neuro-Psychopharmacology and Biological Psychiatry* 34 (2010): 1181–1188.

8: Making Your Brain Smile: Your Brain's Reward System

Adlard, P. A., and Cotman, C. W. "Voluntary Exercise Protects against Stress-Induced Decreases in Brain-Derived Neurotrophic Factor Protein Expression." *Neuroscience* 124 (2004): 985–992.

Aron, A., Fisher, H., Mashek, D. J., Strong, G., Li, H., and Brown, L. L. "Reward, Motivation, and Emotion Systems Associated with Early-Stage Intense Romantic Love." *Journal of Neurophysiology* 94 (2005): 327–337.

Barbour, K. A., Edenfield, T. M., and Blumenthal, J. A. "Exercise as a Treatment for Depression and Other Psychiatric Disorders: A Review." *Journal of Cardiopulmonary Rehabilitation and Prevention* 27 (2007): 359–367.

Bartels, A., and Zeki, S. "The Neural Basis of Romantic Love." *NeuroReport* 11 (2000): 3829–3834.

Berridge, K. C., and Kringelbach, M. L. "Neuroscience of Affect: Brain Mechanisms of Pleasure and Displeasure." *Current Opinion in Neurobiology* 23 (2013): 294–303.

Blumenthal, J. A., Smith, P. J., and Hoffman, B. M. "Is Exercise a Viable Treatment for Depression?" *ACSM's Health and Fitness Journal* 16 (2012): 14–21.

Carroll, M. E., and Lac, S. T. "Autoshaping I.V. Cocaine Self-Administration in Rats: Effects of Nondrug Alternative Reinforcers on Acquisition." *Psychopharmacology* (Berlin) 110 (1993): 5–12.

Carroll, M. E., Lac, S. T., and Nygaard, S. L. "A Concurrently Available Nondrug Reinforcer Prevents the Acquisition or Decreases the

Maintenance of Cocaine-Reinforced Behavior." *Psychopharmacology* (Berlin) 97 (1989): 23–29.

Fontes-Ribeiro, C. A., Marques, E., Pereira, F. C., Silva, A. P., and Macedo, T. R. "May Exercise Prevent Addiction?" *Current Neuropharmacology* 9 (2011): 45–48.

Harbaugh, W. T., Mayr, U., and Burghart, D. R. "Neural Responses to Taxation and Voluntary Giving Reveal Motives for Charitable Donations." *Science* 316 (2007): 1622–1625.

Herring, M. P., O'Connor, P. J., and Dishman, R. K. "The Effect of Exercise Training on Anxiety Symptoms among Patients: A Systematic Review." *Archives of Internal Medicine* 170 (2010):, 321–331.

Hoffman, B. M., Babyak, M. A., Craighead, W. E., Sherwood, A., Doraiswamy, P. M., Coons, M. J., and Blumenthal, J. A. "Exercise and Pharmacotherapy in Patients with Major Depression: One-Year Follow-Up of the SMILE Study." *Psychosomatic Medicine* 73 (2011): 127–133.

Hosseini, M., Alaei, H. A., Naderi, A., Sharifi, M. R., and Zahed, R. "Treadmill Exercise Reduces Self-Administration of Morphine in Male Rats." *Pathophysiology* 16 (2009): 3–7.

Kelley, A. E. "Memory and Addiction: Shared Neural Circuitry and Molecular Mechanisms." *Neuron* 44 (2004): 161–179.

Kringelbach, M. L. "The Human Orbitofrontal Cortex: Linking Reward to Hedonic Experience." *Nature Reviews Neuroscience* 6 (2005): 691–702.

Kringelbach, M. L., and Berridge, K. C. "The Functional Neuroanatomy of Pleasure and Happiness." *Discovery Medicine* 9 (2010): 579–587.

Kringelbach, M. L., and Berridge, K. C. "The Joyful Mind." *Scientific American* 307, no. 2 (2012): 40–45.

Kringelbach, M. L., and Berridge, K. C. "Towards a Functional Neuroanatomy of Pleasure and Happiness." *Trends in Cognitive Science* 13 (2009): 479–487.

Le, M. M., and Koob, G. F. "Drug Addiction: Pathways to the Disease and Pathophysiological Perspectives." *European Neuropsychopharmacology* 17 (2007): 377–393.

Lenoir, M., Serre, F., Cantin, L., and Ahmed, S. H. "Intense Sweetness Surpasses Cocaine Reward." *Public Library of Science* 2 (2007): e698.

Lynch, W. J., Peterson, A. B., Sanchez, V., Abel, J., and Smith, M. A. "Exercise As a Novel Treatment for Drug Addiction: A Neurobiological and Stage-Dependent Hypothesis." *Neuroscience and Biobehavioral Reviews* 37 (2013): 1622–1644.

Mathes, W. F., and Kanarek, R. B. "Persistent Exercise Attenuates

Nicotine—But Not Clonidine-Induced Antinociception in Female Rats." *Pharmacology, Biochemistry, and Behavior* 85 (2006): 762–768.

National Institute on Drug Abuse. *Drugs, Brains, and Behavior: The Science of Addiction* (NIH Publication No. 14-5605). Rockville, MD, 2014.

Nestler, E. J. "The Neurobiology of Cocaine Addiction." *Science and Practice Perspectives* 3 (2005): 4–10.

Olds, J. "Neurophysiology of Drive." *Psychiatric Research Reports* 6 (1956): 15–20.

Olds, J. "A Preliminary Mapping of Electrical Reinforcing Effects in the Rat Brain." *Journal of Comparative Physiological Psychology* 49 (1956): 281–285.

Olds, J. "Runway and Maze Behavior Controlled by Basomedial Forebrain Stimulation in the Rat." *Journal of Comparative Physiological Psychology* 49 (1956): 507–512.

Olds, J., Disterhoft, J. F., Segal, M., Kornblith, C. L., and Hirsh, R. "Learning Centers of Rat Brain Mapped by Measuring Latencies of Conditioned Unit Responses." *Journal of Neurophysiology* 35 (1972): 202–219.

Olds, J., and Milner, P. "Positive Reinforcement Produced by Electrical Stimulation of Septal Area and Other Regions of Rat Brain." *Journal of Comparative Physiological Psychology* 47 (1954): 419–427.

Ortigue, S., Bianchi-Demicheli, F., Hamilton, A. F., and Grafton, S. T. "The Neural Basis of Love As a Subliminal Prime: An Event-Related Functional Magnetic Resonance Imaging Study." *Journal of Cognitive Neuroscience* 19 (2007): 1218–1230.

Perraton, L. G., Kumar, S., and Machotka, Z. "Exercise Parameters in the Treatment of Clinical Depression: A Systematic Review of Randomized Controlled Trials." *Journal of Evaluation in Clinical Practice* 16 (2010): 597–604.

Russo, S. J., Mazei-Robison, M. S., Ables, J. L., and Nestler, E. J. "Neurotrophic Factors and Structural Plasticity in Addiction." *Neuropharmacology* 56, suppl. 1 (2009): 73–82.

Smith, M. A., and Lynch, W. J. "Exercise As a Potential Treatment for Drug Abuse: Evidence from Preclinical Studies." *Frontiers in Psychiatry* 2 (2011): 82.

Smith, M. A., Schmidt, K. T., Iordanou, J. C., and Mustroph, M. L. "Aerobic Exercise Decreases the Positive-Reinforcing Effects of Cocaine." *Drug and Alcohol Dependence* 98 (2008): 129–135.

Taylor, A. H., Ussher, M. H., and Faulkner, G. "The Acute Effects of Exercise on Cigarette Cravings, Withdrawal Symptoms, Affect and Smoking Behaviour: A Systematic Review." *Addiction* 102 (2007): 534–543.

Volkow, N. D., and Wise, R. A. "How Can Drug Addiction Help Us Understand Obesity?" *Nature Neuroscience* 8 (2005): 555–560.

Wise, R. A. "Dopamine, Learning and Motivation." *Nature Reviews Neuroscience* 5 (2004): 483–494.

Xu, X., Aron, A., Brown, L., Cao, G., Feng, T., and Weng, X. "Reward and Motivation Systems: A Brain Mapping Study of Early-Stage Intense Romantic Love in Chinese Participants." *Human Brain Mapping* 32 (2011): 249–257.

Xu, X., Brown, L., Aron, A., Cao, G., Feng, T., Acevedo, B., and Weng, X. "Regional Brain Activity During Early-Stage Intense Romantic Love Predicted Relationship Outcomes after 40 Months: An fMRI Assessment." *Neuroscience Letters* 526 (2012): 33–38.

9: The Creative Brain: Sparking Insight and Divergent Thinking

Berkowitz, A. L., and Ansari, D. "Expertise-Related Deactivation of the Right Temporoparietal Junction during Musical Improvisation." *Neuroimage* 49 (2010): 712–719.

Berkowitz, A. L., and Ansari, D. "Generation of Novel Motor Sequences: The Neural Correlates of Musical Improvisation." *Neuroimage* 41 (2008): 535–543.

Brown, S., Martinez, M. J., and Parsons, L. M. "Music and Language Side by Side in the Brain: A PET Study of the Generation of Melodies and Sentences." *European Journal of Neuroscience* 23 (2006): 2791–2803.

Damasio, A. R. "Toward a Neurobiology of Emotion and Feeling: Operational Concepts and Hypotheses." *The Neuroscientist* 1 (1995): 19–25.

Dietrich, A. "The Cognitive Neuroscience of Creativity." *Psychonomic Bulletin and Review* 11 (2004): 1011–1026.

Dietrich, A., and Kanso, R. "A Review of EEG, ERP, and Neuroimaging Studies of Creativity and Insight." *Psychology Bulletin* 136 (2010): 822–848.

Jung, R. E., Mead, B. S., Carrasco, J., and Flores, R. A. "The Structure of Creative Cognition in the Human Brain." *Frontiers in Human Neuroscience* 7 (2013): 330.

Jung, R. E., Segall, J. M., Jeremy, B. H., Flores, R. A., Smith, S. M., Chavez, R. S., and Haier, R. J. "Neuroanatomy of Creativity." *Human Brain Mapping* 31 (2010): 398–409.

Limb, C. J., and Braun, A. R. "Neural Substrates of Spontaneous Musical Performance: An fMRI Study of Jazz Improvisation." *Public Library of Science* 3 (2008): e1679.

Liu, S., Chow, H. M., Xu, Y., Erkkinen, M. G., Swett, K. E., Eagle, M. W., Rizik-Baer, D. A., and Braun, A. R. "Neural Correlates of Lyrical Improvisation: An fMRI Study of Freestyle Rap." *Scientific Reports* 2 (2012): 834.

Oppezzo, M., and Schwartz, D. L. "Give Your Ideas Some Legs: The Positive Effect of Walking on Creative Thinking." *Journal of Experimental Psychology: Learning, Memory, and Cognition* 40 (2014): 1142–1152.

Seeley, W. W., Matthews, B. R., Crawford, R. K., Gorno-Tempini, M. L., Foti, D., Mackenzie, I. R., and Miller, B. L. "Unravelling *Bolero*: Progressive Aphasia, Transmodal Creativity and the Right Posterior Neocortex." *Brain* 131 (2008): 39–49.

Shamay-Tsoory, S. G., Adler, N., Aharon-Peretz, J., Perry, D., and Mayseless, N. "The Origins of Originality: The Neural Bases of Creative Thinking and Originality." *Neuropsychologia* 49 (2011): 178–185.

10: Meditation and the Brain: Getting Still and Moving It Forward

Fries, P., Nikolic, D., and Singer, W. "The Gamma Cycle." *Trends in Neuroscience* 30(2007): 309–316.

Goldin, P., Ziv, M., Jazaieri, H., Hahn, K., and Gross, J. J. "MBSR vs Aerobic Exercise in Social Anxiety: fMRI of Emotion Regulation of Negative Self-Beliefs." *Social, Cognitive, and Affective Neuroscience* 8 (2013): 65–72.

Holzel, B. K., Carmody, J., Vangel, M., Congleton, C., Yerramsetti, S. M., Gard, T., and Lazar, S. W. "Mindfulness Practice Leads to Increases in Regional Brain Gray Matter Density." *Psychiatry Research* 191 (2011), 36–43.

Holzel, B. K., Ott, U., Gard, T., Hempel, H., Weygandt, M., Morgen, K., and Vaitl, D. "Investigation of Mindfulness Meditation Practitioners with Voxel-Based Morphometry." *Social, Cognitive, and Affective Neuroscience* 3 (2008): 55–61.

Ives-Deliperi, V. L., Solms, M., and Meintjes, E. M. "The Neural Substrates of Mindfulness: An fMRI Investigation." *Social Neuroscience* 6 (2011): 231–242.

Jazaieri, H., Goldin, P. R., Werner, K., Ziv, M., and Gross, J. J. "A Randomized Trial of MBSR Versus Aerobic Exercise for Social Anxiety Disorder." *Journal of Clinical Psychology* 68 (2012): 715–731.

Leung, M. K., Chan, C. C., Yin, J., Lee, C. F., So, K. F., and Lee, T. M. "Increased Gray Matter Volume in the Right Angular and Posterior Parahippocampal Gyri in Loving-Kindness Meditators." *Social, Cognitive, and Affective Neuroscience* 8 (2013): 34–39.

Lutz, A., Greischar, L. L., Rawlings, N. B., Ricard, M., and Davidson, R. J. "Long-Term Meditators Self-Induce High-Amplitude Gamma Synchrony during Mental Practice." *Proceedings of the National Academy of Sciences U.S.A.* 101 (2004): 16369–16373.

Lutz, A., Slagter, H. A., Dunne, J. D., and Davidson, R. J. "Attention Regulation and Monitoring in Meditation." *Trends in Cognitive Science* 12 (2008): 163–169.

MacLean, K. A., Ferrer, E., Aichele, S. R., Bridwell, D. A., Zanesco, A. P., Jacobs, T. L., King, B. G., Rosenberg, E. L., Sahdra, B. K., Shaver, P. R., Wallace, B. A., Mangun, G. R., and Saron, C. D. "Intensive Meditation Training Improves Perceptual Discrimination and Sustained Attention." *Psychological Science* 21 (2010): 829–839.

Singer, W. "Neuronal Synchrony: A Versatile Code for the Definition of Relations?" *Neuron* 24 (1999): 49–25.

Singer, W., and Gray, C. M. "Visual Feature Integration and the Temporal Correlation Hypothesis." *Annual Review of Neuroscience* 18 (1995): 555–586.

Slagter, H. A., Davidson, R. J., and Lutz, A. "Mental Training As a Tool in the Neuroscientific Study of Brain and Cognitive Plasticity." *Frontiers in Human Neuroscience* 5 (2011): 17.

Varela, F., Lachaux, J. P., Rodriguez, E., and Martinerie, J. "The Brainweb: Phase Synchronization and Large-Scale Integration." *Nature Reviews Neuroscience* 2 (2001): 229–239.

INDEX

ABOUT THE AUTHORS

Wendy Suzuki, PhD, runs an interactive research lab at New York University, where her work has been recognized with numerous awards including the prestigious Troland Research Award from the National Academy of Sciences. Her research has focused on understanding the patterns of brain activity underlying long-term memory and, more recently, how aerobic exercise might improve our learning, memory, and cognitive abilities. She is a two-time TEDx speaker and is regularly interviewed on television and in print about her work on the effects of exercise on brain function. She lectures nationally and internationally on her research and serves as a reviewer for many of the top neuroscience journals. She lives in New York City.

www.suzukilab.com

Billie Fitzpatrick has coauthored numerous books, including several *New York Times* bestsellers. She specializes in mind-body health, neuroscience, nutrition, and diet and fitness.

www.billiefitzpatrick.com